Guidance on Water Supply and Sanitation in Extreme Weather Events

Edited by

L Sinisi and R Aertgeerts

Abstract

Extreme weather events, including floods and droughts, occur increasingly often. They affect the operational efficiency and sustainability of water supply, drainage and sewerage infrastructure, and wastewater treatment services, and threaten their protection of public health and the environment. The guidance in this publication summarizes how basic disaster preparedness and early warning procedures can be implemented in the water and wastewater sector, and identifies the specific challenges of extreme weather events to vulnerable areas. It provides advice on the implementation of water safety plans as a tool for risk assessment and management, giving specific attention to small-scale water supply and sanitation systems; and on multisector cooperation, including communication. Finally, based on a review of experience and good practice in the European Region, it summarizes proven adaptation measures for water utilities, drainage and sewerage, and wastewater treatment systems during extreme weather events.

Keywords

WATER SUPPLY
SANITATION
CLIMATE – adverse effects
EMERGENCIES
DISASTER PLANNING
DELIVERY OF HEALTH CARE – organization and administration
RISK MANAGEMENT
GUIDELINES
EUROPE

ISBN 978 92 890 0258 5

Address requests about publications of the WHO Regional Office for Europe to:
    Publications
    WHO Regional Office for Europe
    Scherfigsvej 8
    DK-2100 Copenhagen Ø, Denmark
Alternatively, complete an online request form for documentation, health information, or for permission to quote or translate, on the Regional Office web site (http://www.euro.who.int/pubrequest).

© World Health Organization 2011

All rights reserved. The Regional Office for Europe of the World Health Organization welcomes requests for permission to reproduce or translate its publications, in part or in full.

The designations employed and the presentation of the material in this publication do not imply the expression of any opinion whatsoever on the part of the World Health Organization concerning the legal status of any country, territory, city or area or of its authorities, or concerning the delimitation of its frontiers or boundaries. Dotted lines on maps represent approximate border lines for which there may not yet be full agreement.

The mention of specific companies or of certain manufacturers' products does not imply that they are endorsed or recommended by the World Health Organization in preference to others of a similar nature that are not mentioned. Errors and omissions excepted, the names of proprietary products are distinguished by initial capital letters.

All reasonable precautions have been taken by the World Health Organization to verify the information contained in this publication. However, the published material is being distributed without warranty of any kind, either express or implied. The responsibility for the interpretation and use of the material lies with the reader. In no event shall the World Health Organization be liable for damages arising from its use. The views expressed by authors, editors, or expert groups do not necessarily represent the decisions or the stated policy of the World Health Organization.

Contents

FOREWORD ix

PREFACE xi

ACKNOWLEDGEMENTS xiii

TABLES, FIGURES, CASE STUDIES
AND PICTURES xvii

ACRONYMS AND ABBREVIATIONS xix

GLOSSARY OF TECHNICAL TERMS xx

EXECUTIVE SUMMARY 1

1 EXTREME WEATHER EVENTS AND
WATER SUPPLY AND SANITATION
IN THE EUROPEAN REGION 4
1.1 KEY MESSAGES 5
1.2 INTRODUCTION 6
1.3 EXTREME WEATHER EVENTS:
FACTS AND TRENDS 6
1.4 EXTREMES ARE NOT ONLY DEFINED
BY DIRECT DAMAGE 10
1.5 EXTREMES AND WATER SUPPLY
SANITATION: OLD PROBLEMS,
NEW RISKS AND CHALLENGES 12
1.6 CONCLUSIONS 16

2 BASIC DISASTER PREPAREDNESS
AND EARLY WARNING 18
2.1 KEY MESSAGES 19
2.2 INTRODUCTION 20
2.3 INFORMATION NEEDS: FROM RISK
ASSESSMENT TO RISK REDUCTION . . . 20
2.3.1 Integration of information needs . . 21
2.3.2 Post-event assessment of environmental
and socioeconomic damage 22
2.3.3 Monitoring and forecasting 23

2.4 TOOLS FOR DISASTER PREPAREDNESS
PLANNING 23
2.4.1 Hydrological forecasting tools . . . 23
2.4.2 Early-warning systems 24
2.4.3 Management tools 24
2.4.4 Hazard proofing 26
2.5 ROLE OF THE HEALTH SYSTEM
IN DISASTER PREPAREDNESS
AND EARLY WARNING 26
2.6 CONCLUSION 27

3 COMMUNICATION IN
EXTREME WEATHER EVENTS 28
3.1 KEY MESSAGES 29
3.2 INTRODUCTION: IMPORTANCE OF
A COMMUNICATION STRATEGY 30
3.3 COMMUNICATION ACTIVITIES 30
3.4 PARTNERSHIP IN COMMUNICATION . . 31
3.5 MONITORING AND EVALUATION
OF THE OUTCOMES 31
3.6 CONCLUSIONS 31

4 VULNERABILITY OF COASTAL
AREAS AND BATHING WATERS
IN EXTREME WEATHER EVENTS 32
4.1 KEY MESSAGES 33
4.2 VULNERABILITY OF
INLAND BATHING WATERS 34
4.3 SALINE WATER INTRUSION
IN AQUIFERS USED FOR
THE PRODUCTION OF
DRINKING WATER 35
4.4 CONSEQUENCES OF EXTREME
WEATHER EVENTS FOR
BATHING-WATER QUALITY 37
4.5 WATER QUALITY CHANGES CAUSED
BY EXTREME WEATHER 37

	4.5.1	Stormy rainfalls	37
	4.5.2	Global warming	38
	4.5.3	Droughts and water scarcity	38

4.6 ELEMENTS OF MITIGATION MEASURES FOR BATHING WATERS 39

	4.6.1	Joint information systems and exchange of information	39
	4.6.2	Prevention of stormwater overflow at sewage treatment plants	39
	4.6.3	Prevention of erosion and diffuse pollution by appropriate land-use techniques	39
	4.6.4	Monitoring during extreme weather events and risk assessment	39
	4.6.5	Public awareness and information	39

5 IMPACTS OF CLIMATE CHANGE AND EXTREME EVENTS ON WATERBORNE DISEASES AND HUMAN HEALTH 40

5.1 KEY MESSAGES 41

5.2 LOWER RAINFALL AND DROUGHT 42

5.3 HEAT WAVES 43

5.4 HIGHER WATER TEMPERATURES 43

5.5 COLD SPELLS 43

5.6 HIGHER RAINFALL, MORE INTENSE RAINFALL AND FLOODS 43

5.7 CHANGES IN ECOSYSTEMS 44

5.8 CHANGES IN SEASONALITY 44

5.9 CHANGES IN HUMAN BEHAVIOUR 44

5.10 SLR 44

5.11 CLIMATE CHANGE AND DIARRHOEAL DISEASES 45

5.12 SOME SPECIFIC EXAMPLES OF CLIMATE CHANGE AND WATERBORNE DISEASES 45

6 WATER SAFETY PLANS: AN APPROACH TO MANAGING RISKS ASSOCIATED WITH EXTREME WEATHER EVENTS 50

6.1 KEY MESSAGES 51

6.2 ELEMENTS OF A WSP 52

	6.2.1	WSP team creation and preparatory activities	52
	6.2.2	Description of the water supply system	52
	6.2.3	Identification of hazards and assessment of risks	53
	6.2.4	Determination and validation of control measures and reassessment and prioritization of risks	53
	6.2.5	Development, implementation and maintenance of an improvement plan	53
	6.2.6	Monitoring control measures	53
	6.2.7	Verification of the effectiveness of the WSP	53
	6.2.8	Preparation of management procedures and supporting programmes	55
	6.2.9	Periodic review	55
	6.2.10	IWRM	55

6.3 THE SPECIAL CASE OF SMALL-SCALE WATER SUPPLY SYSTEMS 56

	6.3.1	Importance of small-scale water supply	56
	6.3.2	Challenges in small-scale water supplies	56
	6.3.3	WSPs and small-scale water supplies	58

6.4 WATER SAFETY AND BULK TRANSPORT OF WATER IN EXTREME WEATHER CONDITIONS 58

| | 6.4.1 | Water supply by tanker during drought conditions | 58 |
| | 6.4.2 | Elements of technical guidance for bulk drinking-water transport under drought conditions | 58 |

6.5 GENERAL WSP CHECKLIST 59

7 ADAPTATION MEASURES FOR WATER SUPPLY UTILITIES IN EXTREME WEATHER EVENTS 60

7.1 KEY MESSAGES 61

7.2 VULNERABILITY OF THE WATER CYCLE TO EXTREME WEATHER EVENTS 62

7.3 ADAPTATION MEASURES FOR DROUGHT EVENTS 62

| | 7.3.1 | Adaptation measures in advance of an extreme event – drought | 62 |
| | 7.3.2 | Managing water supplies during extreme events – droughts | 70 |

7.4 ADAPTATION MEASURES FOR FLOOD EVENTS 72

| | 7.4.1 | Adaptation measures in advance of an extreme event – flooding | 72 |

7.5 REGAINING DRINKING-WATER SUPPLY SYSTEMS 75

	7.5.1	Following drought	75
	7.5.2	Following flooding	76
	7.5.3	Disinfecting and restarting domestic distribution systems (house connections and public buildings)	76

7.6 EMERGENCY PLANNING AND INSTITUTIONAL CAPACITY ISSUES 77

	7.6.1	EMERGENCY PLANNING AND PREPAREDNESS 77
	7.6.2	EMERGENCY DISTRIBUTION OF ALTERNATIVE WATER SUPPLIES 78
	7.6.3	INSTITUTIONAL CAPACITY/MUTUAL AID 79
	7.6.4	INTERDEPENDENCIES/ BUSINESS CONTINUITY 79

7.7 SUMMARY 80

8 ADAPTATION MEASURES FOR DRAINAGE, SEWERAGE AND WASTEWATER TREATMENT 82

8.1 KEY MESSAGES 83

8.2 CLIMATE CHANGE IMPACTS ON DRAINAGE SYSTEMS, SEWER SYSTEMS AND WWTPs 84

8.3 ADAPTATION MEASURES TO URBAN WWTPs BEFORE AND DURING DROUGHTS 84

    8.3.1 MAINTENANCE OF SEWER SYSTEMS DURING AN EXTREMELY LONG DRY PERIOD 85

    8.3.2 OPERATION OF UWWTPs DURING EXTREMELY LONG DRY PERIODS – CHANGES IN HYDRAULIC AND POLLUTION LOAD 85

8.4 ADAPTATION MEASURES BEFORE AND DURING FLOODS 85

    8.4.1 CENTRALIZED DRAINAGE/SEWER SYSTEMS AND UWWTPs – PREVENTIVE MEASURES 86

    8.4.2 DECENTRALIZED AND COMMUNITY-BASED SANITATION SYSTEMS – PREVENTIVE MEASURES . . .87

    8.4.3 CENTRALIZED DRAINAGE/SEWER SYSTEMS AND UWWTPs – PROTECTIVE MEASURES DURING FLOODS 87

    8.4.4 DECENTRALIZED AND COMMUNITY-BASED SANITATION SYSTEMS – PROTECTIVE MEASURES DURING FLOODS 88

8.5 RESTORATION OF THE SEWERAGE SYSTEM AND UWWTP 89

    8.5.1 REGAINING AND RESTART OF DRAINAGE/SEWER NETWORK OPERATION 89

    8.5.2 REGAINING AND RESTART OF THE UWWTP OPERATION 89

8.6 SPECIFIC ISSUES OF INDUSTRIAL WWTPs . 90

8.7 SUMMARY 90

8.8 CHECKLIST 91

REFERENCES 97

BIBLIOGRAPHY 103

Foreword

Water and sanitation are key components of any adaptation strategy aimed at preserving human health in a changing world. Extreme weather events such as floods and droughts are occurring with increasing frequency and intensity in the pan-European region. They affect the capacity and operations of existing water and sanitation infrastructures and services, and thereby threaten the protection such services offer to human health and the environment.

Crisis control is not adaptation: while immediate impacts of floods and droughts on water supply and sanitation need to be dealt with on an emergency basis so that people have continued access to safe water and adequate sanitation, long-term policies also need to be developed based on shared experience and scientifically sound evidence.

Water supply and sanitation services have to prepare for the widely anticipated consequences of floods and droughts, or risk compromising access to safe drinking water and adequate sanitation for a substantial number of people in developing and developed countries, with cascading effects on human health, the environment and development. These impacts will also have to be taken into account in the design and construction of new systems, such as new reservoirs.

Parties to the Protocol on Water and Health recognized the importance and the urgency of this issue, and called for the development of a guidance document that would disseminate knowledge and past experience to policy-makers, managers of water supply and sanitation services and the health sector.

The *Guidance on water supply and sanitation in extreme weather events* is a response to this call. It is the result of an extensive consultation process involving experts and institutions from many countries. It recalls the basic scientific findings, provides advice on communication issues, addresses the vulnerability of coastal areas and bathing waters, discusses the impact on human health, places extreme weather events in the context of water safety plans and formulates advice for adaptation measures for water supply and sanitation services during such events.

We hope that the *Guidance* will be used throughout the pan-European region to raise awareness of the importance of these problems, and to support health and environment policy-makers, managers of water supply and sanitation services and civil society at large in formulating resilience assessment in this area and in implementing appropriate programmes of measures.

Ján Kubiš

Executive Secretary,
United Nations Economic Commission for Europe

Stefania Prestigiacomo

Minister of Environment,
Land and Sea, Italy

Zsuzsanna Jakab

WHO
Regional Director for Europe

Preface

It is well known that adverse meteorological events, such as flash floods, droughts, heat waves, cold spells and windstorms are already affecting the European region increasingly effectively, and that adaptation to water cycle variations is the key issue of short- to medium-term strategies in relation to climate change and variability scenarios.

There is widespread consensus relating to the potential for extensive direct damage to people's health and well-being, their assets, and crucial socioeconomic activities (such as agriculture and tourism). Yet there is a lack of knowledge on how to assess the environmental and health effects associated with exposure to the often complex chemical and biological contamination of water and soil that can follow extreme weather.

Water supply and sanitation are crucial determinants of health, especially during emergencies, but failing or compromised water and sanitation services may in themselves pose a risk, a (sometimes irreversible) source of contamination, the impact of which reaches beyond local and national borders.

Under extreme pressure, all the different elements of water utility services, such as abstraction, treatment, mains supply, sewerage systems, sewage treatment works and effluent discharge, all become key environmental health determinants, increasing the risk of chemical and biological contamination of water for human consumption, food and bathing waters, as well as the risk of vector-borne diseases and those spread by rodents. Water and soil contamination will result from effluent discharge during floods, water supplied will show higher concentrations of pollutants in a drought, and the ability of natural ecosystems to assimilate wastes will be affected by inadequate water for sanitation. In large cities, water scarcity will reduce the self-cleaning capacity of sewers, and flooding will exacerbate stormwater overflows and the resulting pollution. Environmental health hazards will be more significant in poor and rural areas, in which utilities infrastructure is lacking, in a poor state, or where small service suppliers cannot cope with adverse weather conditions.

It will not be simply a problem of finding engineering solutions, but a more complex quest for adaptation measures to improve joint coping capacities. The resilience of water supply and sanitation services under changing weather patterns can be significantly challenged, and joint efforts are needed on the part of all sectors involved in the sustainable protection of water resources and the risk management of exposed populations to unhealthy environmental risk factors.

Weather and climate extremes will challenge traditional preventive systems for environmental and health issues, such as environmental monitoring and control, disease surveillance, and early warning in all phases of preparedness, response and recovery. They will also impair the costly sustainability policies needed to preserve safe water. Yet, in practice, their organization, planning and resources are generally structured around old patterns and time series of weather and hydrological data, impairing overall coping capacities for protecting water quality and health.

Adaptation policy should focus on strengthening the capacities of environmental monitoring, early warning and disease surveillance and – importantly – should promote cooperation with relevant stakeholders in risk management, such as utilities managers themselves. This should go beyond compliance with the risks usually identified, to include those associated with climate change. There is a need for dynamic, adaptive measures tailored to local conditions that will consider all risk elements and be able to cope with a wide range of scientific disciplines, as well as with other critical drivers of vulnerability such as land use, urban and rural population and assets, overexploitation of resources, and unsafe use of new sources of water.

Once again it is not only a problem of engineering solutions or financial investments, but of building a comprehensive medium- to long-term risk management approach and consensus.

Water quality will be the major end point of the pressure applied by extreme weather events to water supply and sanitation systems and, at the same time, the starting point of an increased risk of water-related disorders. How much attention is given to this in current adaptation policies?

Many such adaptation policies often focus mainly on long-term water resource management or on improving forecasting abilities, sometimes neglecting the management of new risk elements, such as water supply and sanitation in adverse weather. This sectoral isolation could be a problem: indeed, water services[1] managers – along with environmental health experts – are not usually involved in cooperation and consultation frameworks for medium- to long-term adaptation strategy planning.

With all of this – and more – in mind, under the framework of the 2007–2009 Work Programme of the Protocol on Water and Health of the UNECE (United Nations Economic Commission for Europe) Water Convention,[2] the Italian Ministry of Environment, Land and Sea took up leadership of the Task Force on Extreme Weather Events (established at the Protocol 1st Meeting of Parties), in order to provide *ad hoc* tools to assist countries in the implementation of the provisions of the Protocol in terms of adaptation policies relating to climate change challenges.[3] The Task Force's main mandate was the development of the *Guidance on water supply and sanitation in extreme weather events*.

This *Guidance* is intended to provide an overview of why and how adaptation policies should consider the vulnerability of and new risk elements for health and environment arising from water services management during adverse weather episodes.

Emerging risk factors in conditions of climate variability receive special attention, with a focus on the response capacity of the environment and health sectors, the role of water services managers, and information needs (including a public communication strategy) as key elements of health risk reduction. Special emphasis is given to adaptation measures to ensure safe water supply and sanitation using existing infrastructure.

This document addresses a broad audience, including policy-makers, environment, health and water resources professionals, and water services managers. An integrated environment and health approach steered the development and discussion of the *Guidance*.

The *Guidance* is not intended to be a manual on water supply and sanitation management in emergencies; nor does it aim to be a comprehensive treatise on environment and health risk management in extreme weather conditions. Its aim is broader: by providing an overview of this complex and critical issue, it aims to raise awareness of the need to cope with change that is happening before our eyes – not only climate change, but a new world that plays by new rules, that needs new answers and tools and – above all – needs the motivation to abandon old, ineffective sectoral schemes and approaches.

The multidisciplinary efforts and cooperation that lie behind the development of this *Guidance* can perhaps be considered a pilot exercise for a new way of coping with complex problems.

Last, but not least, I personally feel a strong need to express my gratitude to all my colleagues in the editorial group. In every minute of our meetings, 'phone calls and correspondence I have enjoyed the atmosphere of mutual learning and understanding, the absence of purely sectoral argument or overwhelming attitude, as well as the effort made to build common targets and language. It was quite an outstanding lesson to learn.

I also want to warmly thank all the experts from countries, universities, utility companies, nongovernmental organizations and international organizations that I met during my Task Force experience: their contributions were essential to launching and to improving the development of this *Guidance*. Their encouragement, suggestions and enthusiasm for our task were particularly helpful in dealing with the uncertainties that naturally accompanied this challenging pilot experience.

Luciana Sinisi
Chair
Task Force on Extreme Weather Events

---

1 Paragraph 28 of the European Union Water Framework Directive (WFD) defines water services as "... all services which provide, for households, public institutions or any economic activity: (a) abstraction, impoundment, storage, treatment and distribution of surface water or groundwater, (b) waste-water collection and treatment facilities which subsequently discharge into surface water...".

2 The Protocol on Water and Health to the 1992 Convention on the Protection and Use of Transboundary Watercourses and International Lakes, referred to hereafter as "the Protocol".

3 The main aim of the Protocol is to protect human health and well-being through better water management, including the protection of water ecosystems, and by preventing, controlling and reducing water-related diseases. To meet these goals, its Parties are required to establish national and local targets for the quality of drinking water and the quality of discharges, as well as for the performance of water supply and wastewater treatment. The Parties are also required to reduce outbreaks and the incidence of water-related diseases.

Acknowledgements

The preparation of the *Guidance on water supply and sanitation in extreme weather events* covered a period of four years (2007–2010). It is a direct outcome of the decision of the Meeting of the Parties to the Protocol on Water and Health at its first session to establish the Task Force on Extreme Weather Events, with a mandate to draft the *Guidance*.

The Task Force on Extreme Weather Events, chaired by Italy, met on 21 and 22 April 2008 in Rome. The meeting was hosted by the Italian Ministry for the Environment, Land and Sea. Another meeting was held on 27 and 28 October 2009 at the Palais des Nations, in Geneva, Switzerland. A special workshop was organized in cooperation with the European Union Water Initiative in Bucharest on 25 November 2009, aimed at promoting cooperation with Russian-speaking countries. A small drafting group, which met several times, also played a major role in the preparation of the *Guidance*.

The work was coordinated by Dr Luciana Sinisi – Chair of the Task Force on Extreme Weather Events – from the Italian Higher Institute for Environmental Protection and Research. Valuable support was provided by members of the joint secretariat: mostly by Mr Roger Aertgeerts (WHO Regional Office for Europe) as well as by Ms Francesca Bernardini, Mr Tomasz Juszczak and Ms Ella Behlyarova (United Nations Economic Commission for Europe).

The work of the Task Force involved 75 experts from 23 countries who took active part in its meetings, exchanging information and directly interacting with the Chair. In addition, experts from four United Nations specialized agencies, as well as nongovernmental and water utilities managers' organizations provided significant input.

Special thanks go to Mr Alex Kirby and Ms Nicole Satterly, who edited the publication, and to Ms Giorgia Knechtlin, Ms Lucia Dell'Amura, Ms Olga Carlos and Ms Diana Teeder for their secretarial and administrative support.

The joint secretariat gratefully acknowledges the generous financial and technical support of the Italian Ministry for the Environment, Land and Sea, which made possible the preparation of the *Guidance*, as well as the important contribution by the Ministry of Transport, Public Works and Water Management of the Netherlands.

The joint secretariat particularly thanks the members of the editorial group who drafted and edited the text of the *Guidance*.

Mr Roger Aertgeerts	WHO Regional Office for Europe	Italy
Ms Emma Anakhasyan	Armenian Women for Health and Healthy Environment	Armenia
Ms Benedetta Dell'Anno	Ministry for the Environment, Land and Sea	Italy
Mr Gyula Dura	National Institute of Environmental Health	Hungary
Mr Jim Foster	Drinking-water Inspectorate	United Kingdom
Ms Franziska Matthies	WHO Regional Office for Europe	Italy
Ms Bettina Menne	WHO Regional Office for Europe	Italy
Ms Doubravka Nedvedova	Ministry of the Environment	Czech Republic
Ms Luciana Sinisi	Chair of Task Force Extreme Weather Events, Higher Institute for Environmental Protection and Research	Italy
Mr Giacomo Teruggi	World Meteorological Organization	Switzerland

The joint secretariat also wishes to warmly praise the contributions of the other members of the drafting group.

Mr Bogachan Benli	United Nations Development Programme	Turkey
Mr Roberto Celestini	ACEA ATO2 Roma	Italy
Mr Osvaldo Conio	IRIDE Genova	Italy
Mr Andrea Critto	Euro-Mediterranean Centre for Climate Change	Italy
Ms Alessia Delle Site	ACEA ATO2 Roma	Italy
Ms Helene Di Maggio	Ministry of Environment, Land and Sea	Italy
Mr Emanuele Ferretti	Higher Institute of Public Health	Italy
Mr Enzo Funari	Chair Protocol Task Force Surveillance, Higher Institute of Public Health	Italy
Ms Giuliana Gasparrini	Ministry for the Environment, Land and Sea	Italy
Mr Marco Gatta	FEDERUTILITY	Italy
Ms Lea Kauppi	Finnish Environment Institute	Finland
Mr Azer Khanlarov	Ministry of Emergency Situations	Azerbaijan
Ms Claudia Lasagna	IRIDE Aqua Gas	Italy
Mr Luca Lucentini	Higher Institute of Public Health	Italy
Ms Elena Mauro	FEDERUTILITY	Italy
Ms Lorenza Meucci	SMAT Torino	Italy
Ms Magdalena Mrkvickova	Water Research Institute	Czech Republic
Mr Massimo Ottaviani	Higher Institute of Public Health	Italy
Mr Marco Pelosi	CAP Gestione Milano	Italy
Ms Sabrina Rieti	Higher Institute for Environmental Protection and Research	Italy
Ms Bettina Rickert	Federal Environment Agency	Germany
Mr Ion Shalaru	National Centre for Preventive Medicine	Republic of Moldova
Mr Oliver Schmoll	Federal Environment Agency	Germany
Mr Vaclav Stastny	Water Research Institute	Czech Republic
Mr Jos Timmerman	Ministry of Infrastructure and Environment	Netherlands

The joint secretariat also wishes to express gratitude to the following experts who reviewed the *Guidance* and greatly improved it with their comments and contributions. Their technical and professional input is gratefully acknowledged.

Mr Charles Baubion	World Meteorological Organization	Switzerland
Mr Robert Bos	World Health Organization	Switzerland
Mr Brian Clarke	University of Surrey	United Kingdom
Mr Chee Keong Chew	World Health Organization	Switzerland
Mr Fabio Conti	University of Insubria, Varese	Italy
Ms Jennifer DeFrance	World Health Organization	Switzerland
Mr P J L Dennis	Wessex Water	United Kingdom
Ms Nana Gabriadze	National Centre for Disease Control and Public Health	Georgia
Mr Dominique Gatel	European Federation of National Associations of Drinking-water Suppliers and Wastewater Services	France
Mr Bruce Gordon	World Health Organization	Switzerland
Mr George Kamizoulis	World Health Organization – MEDPOL	Greece
Mr Alexander Mindorashvili	Ministry of Environment Protection and Natural Resources	Georgia
Mr Paul-Cristian Ionescu	Ministry of Health	Romania
Mr Bruce Rhodes	Melbourne Water	Australia
Ms Sabrina Sorlini	University of Brescia	Italy
Mr Sandro Teruggi	Ecostudio	Italy
Ms Leylakhanim Taghizade	Republican Centre for Hygiene and Epidemiology	Azerbaijan
Ms Sinead Tuite	World Health Organization	Switzerland
Mr Michiel Van Peteghem	Flemish Environment Agency	Belgium

The joint secretariat also wishes to thank all participants of Task Force on Extreme Weather Events meetings for their comments and contributions to the overall process.

Ms Loreta Asokliene	Ministry of Health	Lithuania
Ms Martina Behanova	Public Health Authority	Slovakia
Ms Gabriella Ceci	Mediterranean Centre for Climate Change	Italy
Mr Massimo Cozzone	Ministry for the Environment, Land and Sea	Italy
Ms Francesca De Maio	Higher Institute for Environmental Protection and Research	Italy
Mr Kemal Dokuyucu	Turkish Meteorological Service	Turkey
Ms Zsuzsanna Engi	West-Transdanubian District Environmental and Water Authority	Hungary
Ms Franziska Matthies	WHO Regional Office for Europe	Italy
Ms Judith Plutzer	National Institute of Environmental Health	Hungary
Mr Valery Filonau	Republican Scientific – Practical Centre of Hygiene	Belarus
Ms Svitlana Gariyenchyck	Ministry of Environment Protection	Ukraine
Mr Vladimir Garaba	Environmental Movement	Republic of Moldova
Ms Ghazaros Hakobyan	State Hygiene and Anti-epidemic Inspectorate	Armenia
Mr Nazmi Kagniciogly	General Directorate of State Hydraulic Works	Turkey
Mr Merab Kandelaki	Statskalkanali	Georgia
Ms Zdenka Kelnarova	Ministry of the Environment	Slovakia
Mr Ahmad Mamadov	Sukanal Scientific Research Institute	Azerbaijan
Mr Viacheslav Manukalo	State Hydrometeorological Service	Ukraine
Ms Taisa Neronova	State Agency on Environment Protection and Forestry	Kyrgyzstan
Mr Erkki Santala	Finnish Environment Institute	Finland
Mr Pierre Studer	Swiss Office of Public Health	Switzerland
Ms Aijamal Tleulessova	Balhash-Alacol Basin Water-Economic Board	Kazakhstan
Mr Ayhan Taskin	General Directorate of State Hydraulic Works	Turkey
Ms Jessica Tuscano	Higher Institute for Environmental Protection and Research	Italy
Mr Asif Verdiyev	Ministry of Ecology and Natural Resources	Azerbaijan
Ms Tanja Wolf	WHO Regional Office for Europe	Italy
Mr Boril Zadneprovski	Ministry of Environment and Water	Bulgaria

Tables, Figures, Case Studies and Pictures

Tables

Table 1	Projected climate change impacts.	7
Table 2	Central Europe's most costly flood catastrophes, 1993–2006	9
Table 3	Minimum losses from agricultural drought and cost of relief operations in central Asia and the Caucasus, 2000–2001	10
Table 4	Selection of major windstorm catastrophes in central Europe, 1990–2007	11
Table 5	Pre-vulnerability assessment	22
Table 6	Data needs for integrated assessment	23
Table 7	Hydrological forecasting tools	24
Table 8	Health system planning for flood preparedness	27
Table 9	Classification of the impact of climate change on the vulnerability of coastal waters according to the DPSIR approach	34
Table 10	Classification of the impact of climate change on the vulnerability of inland bathing water according to the DPSIR approach	35
Table 11	Classification of saline irrigation water	36
Table 12	Observed and projected changes in climate conditions: potential risks and opportunities	42
Table 13	Projected incident cases under high and low emission scenarios by 2030	46
Table 14	Summary table on pathogens and health significance	47
Table 15	Typical hazards associated with extreme weather events	54
Table 16	Access to improved drinking-water sources in rural areas in the European region	57
Table 17	Potential impact on feature or system in water supply and sanitation	64
Table 18	Examples of adaptation measures	67
Table 19	Examples of proactive measures	68
Table 20	Water treatment works adaptation	69
Table 21	Adaptation of distribution systems	70
Table 22	Options for demand management	71
Table 23	Examples of proactive adaptation measures	73
Table 24	Adaptation activities (floods)	74
Table 25	Key principles in recovering a water supply system (summary table)	76
Table 26	Impacts and mitigation measures	79
Table 27	Checklist for adaptation measures for drainage and sewerage systems	93

Figures

Fig. 1	Number of extreme weather disasters in the UNECE region and globally, 1980–2008	8
Fig. 2	Number of people affected by extreme weather disasters in the UNECE region, 1970–2008	8
Fig. 3	River catchments affected by flooding, 1998–2005	13
Fig. 4	Concept framework for the assessment of drought impacts	13
Fig. 5	Percentage of the population with domestic connection to improved sanitation facilities in urban and rural areas, 2006	14
Fig. 6	Disaster management process	20
Fig. 7	Components of risk	20
Fig. 8	Risk map developed using Geographic Information System	21
Fig. 9	Four elements of people-centred early-warning systems	25
Fig. 10	Schematic illustration of a river catchment	63
Fig. 11	Intervention options in extreme events (floods)	71
Fig. 12	Flood protection – order of intervention	74
Fig. 13	Differentiated water quality requirements	78

Case Studies

Case study 1	Consequences of drought in a shallow lake (Lake Balaton), Hungary, 2003	37
Case study 2	Cold spell in Tajikistan, 2008	43
Case study 3	Environmental health aspects of flooded karstic drinking-water resources, Hungary	45
Case study 4	Changes in the marine food web in Europe	45
Case study 5	Roof-top rainwater harvesting in a semi-arid climate	67
Case study 6	Impact of climate change on water resources in Azerbaijan	67
Case study 7	Unprecedented cyanobacterial bloom and MC production in a drinking-water reservoir in the south of Italy	68
Case study 8	Impact of water supply and usage improvement, Turkey	70
Case study 9	Transboundary transfer of raw water resources in Azerbaijan	72
Case study 10	Recovering a water supply system after floods, England, 2007	77
Case study 11	Water supply tanks disinfection	79
Case study 12	Water supply problems in the case of power cuts caused by extreme weather conditions, Hungary	80
Case study 13	Scenarios of water entering UWWTP areas during flood, Czech Republic, 2002	88
Case study 14	Damage caused by the flooding of the IWWTP in Rozoky, Czech Republic, 2002	91
Case study 15	Sewerage network and sanitation planning, management and recovery in case of extreme events, Belgium	92

Pictures

Picture 1	Flooding of the river Bóda, Hungary, 2003	4
Picture 2	Aftermath of a flooding event, Tbilisi, Georgia, 2009	18
Picture 3	Handling media skilfully is essential	28
Picture 4	Bathers being forced out of Lake Balaton, Hungary, by a sudden summer storm	32
Picture 5	Algal contamination in Lake Balaton	38
Picture 6	Health workers in front of a hospital in Muynak, Karakalpakstan in March 2008	40
Picture 7	Water safety plans control risks from source to tap, through the different control points in water treatment plants	50
Picture 8	Flooded drinking-water treatment plant, Mythe, Gloucestershire (United Kingdom), 2007	60
Picture 9	Flooding of urban wastewater treatment plant, Prague, Czech Republic (2002)	82

Acronyms and Abbreviations

ADPC	Asian Disaster Preparedness Centre
ADRC	Asian Disaster Reduction Centre, Japan
APFM	Associated Programme on Flood Management
BAT	best available technology
BOD	biochemical oxygen demand
CIMO	Commission for Instruments and Methods for Observation
CRED	Centre for Research on the Epidemiology of Disasters
CSO	combined sewer overflow
DFID	Department for International Development, United Kingdom
DPSIR	Drivers, Pressures, State, Impacts, Responses
DTI	daily tolerable intake
DWD	Drinking-water Directive
EEA	European Environment Agency
EECCA	Eastern Europe, Caucasus and central Asia
EM-DAT	International Disaster Database (of CRED)
ENHIS	Environmental Health Information System (of WHO)
EUREAU	European Federation of National Associations of Drinking-water Suppliers and Waste Water Services
EU	European Union
EUWI	European Union Water Initiative
GAC	granulated activated carbon
GCM	global climate model
GDP	gross domestic product
HAB	harmful algal Blooms
GIS	geographic information system
HAV	viral hepatitis A
HVAC	heating, ventilation and air conditioning
IFM	Integrated Flood Management
IPCC	Intergovernmental Panel on Climate Change
ISOP	incident situation operational procedure
ISPRA	Higher Institute for Environmental Protection and Research, Italy
IWA	International Water Association
IWRM	Integrated Water Resources Management
IWWTP	industrial wastewater treatment plant
JMP	WHO/UNICEF Joint Monitoring Programme
JRC	Joint Research Centre (ISPRA)
MCs	microcystin
NHMS	national hydro/meteorological services
NODs	nodularins
NOOA	National Oceanic and Atmospheric Administration, United States
NSICD	National Snow and Ice Data Centre
OECD	Organisation for Economic Co-operation and Development
PAHO	Pan-American Health Organization
PESETA	Projection of Economic impacts of climate change in Sectors of the European Union based on bottom-up Analysis
PSP	paralytic shellfish poisoning
RASFF	Rapid Alert System for Food and Feed (of the European Commission)
RCC	regional climate centre
RCOF	regional climate outlook forum
RTC	real-time control
SCADA	Supervisory Control and Data Acquisition
SEE	south-eastern Europe
SLR	sea level rise
SOP	standard operating procedures
STW	sewage treatment works
TDS	total dissolved solids
TFEWE	Task Force on Extreme Weather Events
UNDAC	United Nations Disaster Assessment and Coordination
UNECE	United Nations Economic Commission for Europe
UNHCR	Office of the United Nations High Commissioner for Refugees
UNISDR	United Nations International Strategy for Disaster Reduction
UNOCHA	United Nations Office for the Coordination of Humanitarian Affairs
USDA	United States Department of Agriculture
UWWD	Urban Waste Water Directive
UWWTP	Urban wastewater treatment plant
VMM	Flemish Environment Agency
WBCSD	World Business Council on Sustainable Development
WMO	World Meteorological Organization
WHO	World Health Organization
WFD	Water Framework Directive
WSP	water safety plan
WTW	water treatment works
WWTP	wastewater treatment plant

Glossary of Technical Terms

Term	Meaning
Ad hoc information	Information not gathered as a result of rigorously planned scientific experiments but obtained, for example, during the daily operations of a water supply system
Adduction	The infrastructure of works that brings water to the treatment plants
Autotroph/ autotrophic organism	An organism that uses carbon dioxide as a source of the carbon it needs for building new cells. Autotrophs do not, like heterotrophs, need organic carbon; they can consume dissolved nitrates or ammonium salts and they include nitrifying bacteria and algae
Backwater	The body of relatively still water in coves or covering low-lying areas and having access to the main body of water"-(*Source:* The Drinking Water Dictionary, AWWA).
Computation	The use of computers, especially as a subject of research or study
Commensal	Relating to an association between two organisms, in which one benefits and the other derives neither benefit nor harm
Conductivity	The ability of a conductor to pass electrical current. For water, the value in µS is roughly proportional to the concentration of dissolved solids, i.e. 150 µS/cm corresponds to about 100 mg/l of totally dissolved solids
Congener	Belonging to the same genus
Diatoms	The common name for a member of the phylum Bacillariophyta; a group of algae characterized by delicately marked thin double shells of silica
Dejection	Usually a solid, as opposed to manure, for example, which can have both solid and liquid elements (to deject = to throw down)
Empower	To give authorization and power to, but also give strength and confidence to
Externality	Something that influences a process but is not regulated by it, e.g. cattle grazing next to a well (negative externality).
Incidence	The occurrence, rate or frequency of a disease
Inflow	Water that flows into a pipe
Karst	A landscape underlain with rock formations that have been eroded by dissolution, producing fissures and sinkholes through which pollutants can enter and contaminate the underlying aquifers
Limnology	The study of inland fresh waters, especially lakes and ponds (adaptation from Britannica online)
Organoleptic	Relating to sensory organs; parameters and properties of a substance that can affect the human senses or organs
Outbreaks	Time-limited, usually high-casualty events
Passable	Able to be travelled along or on
Percentile	1 each of 100 equal groups into which a population can be divided according to the distribution of values of a particular variable
Pervious	Able to allow water to pass through, permeable
Phenology	The study of cyclic and seasonal natural phenomena, especially in relation to climate and plant and animal life
Plumbosolvency	The dissolution of lead in water
Polishing pond (or maturation pond)	An aerobic waste stabilization pond, usually following a facultative pond, typically 1–3 m deep, usually used for tertiary treatment of wastewater. Typical retention time approximately 2 days, organic load 100 kg biochemical oxygen demand (BOD) per ha per day, for an effluent quality of 25 mg/l or lower

Population equivalent	The organic strength of wastewater, expressed as equivalent population based on the mass of BOD_5^{20} that one person excretes in one day
Protagonism	Efforts to promote one sector over another
Protection zone	Area of limited activity aimed at avoiding contamination of aquatic resources by anthropogenic activity
Protocol	Here: the accepted code of behaviour in a given circumstance. Protocols can vary from water service to water service
Residual chlorine	Chlorine that exists in water either as hypochlorite (free residual chlorine) or in combination with other substances – particularly ammonia – as combined residual chlorine. The term "residual" refers to the process whereby water contains X mg/l chlorine when it leaves a treatment plant, of which Y mg/l is consumed during transport. The amount remaining at the point of consumption is the residual chlorine
Reticulated	Resembling a net or network, especially having veins, fibres or lines crossing, e.g. a network of pipes distributing water in a city
Solute	That which is dissolved, for example kitchen salt in water
Stratification	Organized or formed in strata (layers)
Symbiont	An organism living in symbiosis with another; the close and usually obligatory association of two organisms of different species living together, not necessarily to their mutual benefit. The term "symbiosis" is often used exclusively for an association in which both partners benefit, which is more properly called "mutualism"
Unconfined (groundwater)	Unconfined groundwater is groundwater that is overlaid by an unsaturated zone, as opposed to "confined" groundwater, which is delimited by a saturated zone (or by an impermeable layer, usually clay or granite)
Vector	A person, animal, or plant which carries a pathogenic agent and acts as a potential source of infection for members of another species
Water matrix	Water and its dissolved or suspended constituents

Executive Summary

Without any doubt, water supply and sanitation, together with energy, have contributed to a major change of living conditions in human history.

There is also wide agreement on the importance of water supply and sanitation systems to environmental issues and health problems, social services, poverty alleviation, sustainable water resources management, food production and security, drinking-water supply and water-related natural disasters.

When the weather is abnormal or the climate is under pressure, water and wastewater services systems stand to lose much of their environment and health benefits, for two main reasons:

- they lose their ability to deliver the services required because of direct infrastructure damage (from floods, windstorms and tide surges) or from lack of water (e.g. when a cold spell turns water to ice);

- they become a significant source of chemical and biological contamination of ecosystems, water bodies and soil by means of their discharges and polluted overload.

This contamination may sometimes be irreversible, and may also affect areas beyond local and national borders. This transboundary aspect is particularly important in the European region, in which there are over 150 transboundary rivers whose combined watersheds cover more than 40% of the land surface area of the region (UNECE, 2009b).

Then, when adverse weather events recur, or when water cycle variations show patterns that do not match usual meteorological and hydrological sequences, there is a serious threat to sustainable livelihoods and to the health of exposed populations.

Such events have already occurred in the European region in the past 20 years. The average number of annual disastrous weather- and climate-related events in Europe increased by about 65% between 1998 and 2007 (EEA, 2008). Overall losses caused by such events increased during the period 1980–2007 from a previous decadal average of less than €7.2 billion (1980–1989) to about €13.7 billion (1998–2007).

In terms of social impacts, the CRED Emergency Events Database (EM-DAT, 2009) (Centre for Research on the Epidemiology of Disasters) shows that in the past 20 years, about 40 million people required health assistance and had basic survival needs, such as safe shelter, medical assistance, safe water supply and sanitation. This represented an increase of about 400% compared to the 8 million people affected in the previous two decades (1970–1990).

According to a 2006 World Bank report (World Bank, 2006), all countries of central Asia and the Caucasus are highly exposed to meteorological and hydrological drought.

A severe and widespread drought in 2000–2001 wiped out between 10% and 26% of crop and livestock production in Armenia, Georgia, and Tajikistan (3–6% of overall gross domestic product (GDP)).

In the same period the hardest-hit communities of Armenia, Azerbaijan, Georgia, Tajikistan and Uzbekistan required food, drinking-water, and agricultural relief supplies costing about US$ 190 million.

Windstorm loss data across 29 European countries also show an increase of more than 200% in the past 20 years, compared with the 1970–1989 period.[4]

These extremes will amplify the existing vulnerabilities of water supply and sanitation systems in the region. Indeed, at the European Union (EU) level more than 12 million citizens lack safe sanitation and those in rural areas are still very vulnerable compared to urban populations. Until 2005 in many countries of the WHO European Region the percentage of the population connected to wastewater treatment facilities ranged from between 15% and under 50% (WHO Regional Office for Europe, 2009). The EU Urban Waste Water Directive (UWWD) (Council of the European Union, 1991) dates back to 1991 and there are currently no plans to revise it. It is probably the most expensive Directive ever adopted, with investment costs of about €30 billion for the 12 newer Member States (that have joined the EU since May 2004) (Buitenkamp & Stintzing, 2008). Yet, not all Member States (EEA, 2005a; BIPE, 2006) are in compliance with all of its provisions.

Environment and health risk management should cope under extreme conditions with a wide range of different scientific and other information, as well as with the internal vulnerabilities of these services, such as pressure on existing networks, quality of performance in critical conditions, implementation of technology development and safe delivery of services.

Resilience requirements aside, unsuitable management of the infrastructure may also have negative impacts on general water management and may, in turn, affect healthy water and wastewater services.

How much of this will affect the tasks and objectives of the Protocol on Water and Health[5], particularly the protection of human health and well-being by better water management, including the protection of water ecosystems, and the prevention, control and reduction of waterborne outbreaks and water-related diseases?

What could be the added value for Protocol goals in addressing water supply and sanitation issues under the conditions likely to prevail during extreme weather events?

How would the overall United Nations Economic Commission for Europe (UNECE) Water Convention adaptation strategies and initiatives benefit from such an exercise?[6]

With these demanding questions in mind, the multidisciplinary editorial and drafting group established under the framework of the Protocol Task Force on Extreme Weather Events coped with the challenges of designing, elaborating and writing this *Guidance on water supply and sanitation in extreme weather events*. They were led by the following overarching considerations.

- In extreme conditions, water supply and sanitation are crucial determinants of health, both because of the need for safe services in emergencies and because they are themselves a significant risk factor as potential sources of heavy pollution.

- In major adverse weather events such as floods and droughts, there is not only direct damage to health and for society at large; health hazards also derive from the increased risk of chemical and biological contamination of water for human consumption, food production and consumption, and bathing waters, as well as risk of changes in the distribution of disease-carrying vectors and rodents.

- Changes in water quality and quantity can be considered as the major environmental end point of the pressure exerted by extreme events and, at the same time, water cycle changes can be seen as the starting point of unhealthy environmental conditions. The efficiency of water supply and sanitation systems plays a major role in this.

- A focused overall environment and health risk management approach needs at first to cope with a wide range of science (engineering, operational, developmental, financial and management concepts), institutional stakeholders (utilities, land users, water resource managers) and frameworks (Integrated Water Resources Management (IWRM)), disaster reduction strategies, sustainable development, flood and drought risk management, early warning and forecasting).

- Elements of the principal infrastructure components required to satisfy water cycle management objectives (such as network of drinkable water supply; sewage collection, treatment and effluent disposal; stormwater collection, treatment and disposal; as well as reclaimed (recycled) water collection, storage, treatment and re-use or disposal) are affected in different ways.

(The *Guidance* is also intended to address adaptation measures for all these infrastructure components to benefit investment planning.)

- Any risk management strategies would benefit from other crucial tools, such as adequate information tools and communication strategies and the strengthening of the coping capacities of environmental monitoring, early warning and disease surveillance.

- Current challenges experienced by utilities managers derive both from climate-related factors and those not related to climate. This requires a more integrated and complex approach.

Many drivers not related to climate act together with global changes to compound and affect extreme events, producing vulnerabilities in hydrological and ecosystems, as well as in economic and social systems.

---

4 JI Barredo (JRC, ISPRA), personal communication, 2009.
5 See Preface.
6 The Protocol Task Force on Extreme Weather Events cooperates on health, water and extreme weather-related issues, contributing to the development of the UNECE *Guidance on water and adaptation to climate change*, prepared by the UNECE Water Convention Task Force on Water and Climate (UNECE, 2009a).

Land-use changes play a role in the rainfall–runoff relationship. Deforestation, urbanization and the reduction of wetlands impair available water storage capacity and increase the runoff coefficient, leading to a growth in flood amplitude and reduction of the time-to-peak. Urbanization has adversely influenced flood hazard by increasing the number of sealed areas and infrastructures. The trend towards growing urbanization is also leading to unplanned slum neighbourhoods with poor or non-existent basic water and wastewater services. Unsound dwellings, unsafe water for drinking and hygiene, overcrowding and a lack of basic infrastructure such as sewerage networks characterize many informal settlements.

In some countries, market-oriented policies designed to facilitate crop production through huge changes in river basins resulted in impaired drinking-water supplies to local populations.

The increase in water pricing is leading poor people, especially in small communities and rural areas, to use old, unsafe wells and unsafe new sources, such as untreated recycled water.

In the *Guidance* development the *ad hoc* editorial and drafting group faced all the aforementioned challenges. Problem analysis, together with the professional experience of experts from various countries and international organizations, and the involvement of utilities managers were crucial to developing the draft of the document.

The design and outlook of the *Guidance* is intended to provide an overview of why and how adaptation policies should consider the new vulnerabilities and risks for health and the environment which arise from water and waste services management in adverse weather events. Major topics addressed include those listed here.

- In Chapter 1 (Extreme weather events and water supply and sanitation in the European region), together with an overview of current data showing how climate change and variability is already increasingly hitting the European region, an overview of water supply and sanitation management system vulnerabilities in the region is provided, both relating to climate and otherwise. The principal aim is to raise awareness and foster preparedness among decision-makers and stakeholders to ensure the adequate planning of adaptation measures for water supply and sanitation systems, and to enhance the abilities of all actors and sectors involved in risk management, such as early-warning systems and the environmental and health sectors. Discussion also stressed the crucial role played by managers of the entire water cycle, the need to involve them in planning adaptation policies, the urgent challenges of climate change for an industry facing the need to invest in technology, new facilities and staff training, and the potential for conflicts with mitigation policies.

- Chapter 2 (Basic disaster preparedness and early warning) recalls key elements of available information tools needed for monitoring, forecasting and vulnerability assessment to support risk reduction strategies and disaster preparedness. This chapter also includes a reminder of the role of health services in preventing health risks.

- Guidance is given in Chapter 3 (Communication in extreme weather events) on communication strategies as an integral part of adaptation and risk prevention: how to properly communicate risk to people, as well as how to build and deliver messages to the public.

- The special vulnerability of coastal areas (both inland and marine waters) to climate change and extreme events – along with the need to devise specific environmental approaches (of health relevance) to support focused adaptation measures – is addressed in Chapter 4 (Vulnerability of coastal areas and bathing waters in extreme weather events).

- Human health is dramatically impacted by extreme weather events such as floods and droughts, but also by changes in the hydrological cycle and the disruption of ecosystem services. These relations and specifically the changing threat of water-related diseases are covered in Chapter 5 (Impacts of climate change and extreme events on waterborne diseases and human health).

- Water safety plans (WSPs) – and the general risk assessment/risk management approach that ensures the safety of water from resource to tap – are valid approaches to managing risks associated with extreme weather events. This is demonstrated in Chapter 6 (WSPs: an approach to managing risks associated with extreme weather events).

- An overview of the health risks and impacts associated with floods, droughts, cold spells and heat waves, as well as their relationship with water safety and the preparedness challenges for the health, environment and water sectors are discussed in Chapter 7 (Adaptation measures for water supply utilities in extreme weather events).

- Chapter 8 discusses (adaptation measures for drainage, sewerage and wastewater treatment).

An integrated environment and health approach steers the overall document. Possible cross-cutting issues, such as the role of environment, climate and health sectors in weather extremes; the need for policy dialogue and multisectoral partnership building; the challenge of different settings (urban versus rural; small versus centralized, large-scale suppliers) are addressed in all topics, although they would need a more extensive analysis outside the goals of the *Guidance*.

The *Guidance* obviously aims to be neither a complete manual of water supply and sanitation management in emergencies, nor a comprehensive guide to environment and health risk management in extremes in the emergency and post-event recovery phase.

The aim is broader: providing an overview of the complex and critical issues is intended to raise awareness of the need to use available institutional and technical instruments to cope with change that is taking place: not only climate change, but a whole new world that plays by new rules, that needs new tools, new answers and, above all, the motivation to abandon old, ineffective sectoral schemes and approaches to preserving healthy waters.

CHAPTER 1

EXTREME WEATHER EVENTS AND WATER SUPPLY AND SANITATION IN THE EUROPEAN REGION

Luciana Sinisi
Chair, Task Force on Extreme Weather Events
Protocol on Water and Health

Flooding of the river Bóda, Hungary, 2003
© National Institute of Environmental Health, Budapest (Hungary)/Gyula Dura

Extreme Weather Events and Water Supply and Sanitation in the European Region

1.1 Key Messages

Extreme weather events have already heavily affected the European region with growing frequency in the past 20 years, in line with the global worldwide trend.

WHO estimates that diarrhoeal disease has caused over 13 500 deaths in those aged under 14 years in the eastern European and central Asian countries of the European region, with a strong association with poor drinking-water quality and hygiene and a lack of sewerage and sanitation (Valent et al., 2004).

Since 1990 over 450 floods and more than 300 heavy windstorms in the region have been classified as disasters. About 40 million people were left in need of basic survival requirements, such as food, water, shelter, sanitation and immediate medical assistance.

Although there is wide agreement on and awareness of direct damage to societies and people's health, there is a lack of knowledge about assessing environmental concerns and health effects associated with exposure to the complex contamination of water bodies and soil that follows extreme weather events.

Overall losses resulting from weather- and climate-related events have clearly increased during the past 20 years in EU countries.

Under severe weather conditions, water and wastewater services are no longer beneficial delivery services, but a significant source of chemical and biological contamination. Sometimes this is irreversible and reaches beyond local and national borders.

In the past decade, European countries' major river catchments have experienced several flood episodes, and a higher-than-average rate of global sea level rise (SLR) (of approximately 3.1 mm/year) has been estimated for the past 15 years.

Infrastructure elements of water supply and sanitation systems show specific, individual vulnerabilities to different types of extremes.

Drought is heavily affecting central Asia and the Caucasus. In the 2000–2001 drought period the hardest-hit communities of Armenia, Azerbaijan, Georgia, Tajikistan and Uzbekistan required food, drinking-water, and agricultural relief supplies, costing about US$190 million.

Adaptation and environment and health risk reduction strategies could take into consideration the management of new risk elements for water safety and health hazards associated with poor performance of water supply and sanitation in extremes, in both the short and medium terms.

1.2 Introduction

The water cycle is the main mechanism governing our weather and climate. Changes in climate variability and in the frequency and patterns of extreme weather events in recent decades across the European region are plain to see. No-regret adaptation measures (meaning options that would be justified by their benefits even in the absence of any strong link with man-made climate change) are urgently needed to cope with the harm to environmental resources, ecosystems, sustainable livelihoods and people's health (Table 1. Projected climate change impacts).

Climate variability is also already being observed and it is expected to increase in most locations worldwide.

The term "severe weather" is often used interchangeably with "extreme weather" or "extreme weather events", even though they have distinct meanings: literally, extreme weather events are events that are rare within their statistical reference distribution at a particular place (Parry et al., 2007); while severe weather refers to any dangerous meteorological or hydrometeorological phenomenon, of varying duration, which risks causing major damage, serious social disruption and loss of human life (WMO, 2005).[7]

The Intergovernmental Panel on Climate Change (IPCC) in its fourth assessment report (Parry et al., 2007; Confalonieri et al., 2007) provides scenarios relating to changes in weather events and an overview of projected impacts on specific sectors. These are summarized in Table 1.

Extreme events have a number of attributes that contribute to them being multifaceted phenomena. These include their rate (frequency), intensity, volatility (shape) and dependence (clustering in space or time). A range of hypotheses exists concerning how extreme events might alter with climate change. These include the hypotheses "no change", "mean effect" (increase in mean but not variability), "variance effect" (increase in range) and "structural change" (increase in mean and skew of low-probability events).

Many organizations (e.g. World Meteorological Organization (WMO), National Oceanic and Atmospheric Administration (NOOA)) recognize a link between global warming and the increase of extremes.

However, the devastating impacts of increasing numbers of extreme weather events worldwide have already promoted several international framework and interagency programmes (e.g. the United Nations HYOGO Framework for Action 2005–2015) (UNISDR, 2005) in order to promote disaster resilience and the introduction of risk reduction strategies into policies of adaptation to climate change. WHO and UNECE have also been deeply involved in several initiatives and projects relating to climate change and extreme events challenges, and many countries have launched adaptation strategies.

Yet in practice many efforts have been undertaken to improve IWRM and early warning capacities, while neglecting the fact that in extreme conditions water supply and sanitation are crucial determinants of health, both because of the need for safe services in emergencies and because they are themselves a significant risk factor as potential sources of heavy pollution for water and soil. It is no longer only a question of engineering or financial solutions; as a matter of urgency an overall assessment is required of new risks for water safety and health hazards associated with the poor performance of water supply and sanitation in extreme situations.

The aim of this chapter is to briefly review the issue of trends and impacts of extremes in the European region in order to raise the awareness of decision-makers and relevant stakeholders regarding the need to re-assess the tools and capacities to cope with such extremes. The chapter also introduces the topic of the role of water supply and sanitation performance as a major environmental health determinant for water safety.

1.3 Extreme Weather Events: Facts and Trends

Extreme weather events such as floods or windstorms are the most frequent natural disasters (as defined by the international disaster database (EM-DAT) criteria) observed in the last century. Their frequency has shown a striking worldwide increasing trend in the last two decades, as well as across the UNECE region (European region plus United States and Canada). This is shown in Fig. 1 (Number of extreme weather disasters in the UNECE region and globally, 1980–2008).

Fig. 2 (Number of people affected by extreme weather disasters in the UNECE region, 1970–2008) shows the number of people affected by disasters, defined as "people requiring immediate assistance during a period of emergency, i.e. requiring basic survival needs such as food, water, shelter, sanitation and immediate medical assistance". The data show an overall increase by about 400% in the 1970–1989 period (38 million compared to 8 million), although a slight decline is noticeable in the last decade, possibly due to improved emergency response systems.

However, these numbers describe simply the tip of the iceberg, since they relate only to disaster conditions. The European region lacks a comprehensive database of significant adverse weather events that – even if they do not match EM-DAT's disaster criteria – will impose many environment and health

---

[7] "Severe weather event: a meteorological or hydrometeorological event that presents a risk of adverse impact to life, property or national infrastructure, on any geophysical scale and timescale of a few weeks or less, and that requires action to both communicate to the public and to the responsible authorities, and to reduce the impact." (WMO, 2005).

Table 1. Projected climate change impacts

Projected change	Projected impacts by sector			
	Agriculture, forestry	Water resources	Human health/mortality	Industry/settlement/society
Warmer/fewer cold days/nights; warmer/more hot days/nights over most land areas	Increased yields in colder environments; decreased yields in warmer environments	Effects on water resources relying on snow melt	Reduced human mortality from decreased cold exposure	Reduced energy demand for heating; increased demand for cooling; declining air quality in cities; reduced effects of snow, ice, etc.
Warm spells/heat waves: frequency increases over most land areas	Reduced yields in warmer regions due to heat stress at key development stages; fire danger increase	Increased water demand; water quality problems, e.g. algal blooms	Increased risk of heat-related mortality	Reduction in quality of life for people in warm areas without air conditioning; impacts on elderly and very young; reduced thermal power production efficiency
Heavy precipitation events: frequency increases over most areas	Damage to crops; soil erosion, inability to cultivate land, water logging of soils	Adverse effects on quality of surface and groundwater; contamination of water supply	Deaths, injuries, infectious diseases, allergies and dermatitis from floods and landslides	Disruption of settlements, commerce, transport and societies due to flooding; pressures on urban and rural infrastructures
Area affected by drought: increases	Land degradation, lower yields/crop damage and failure; livestock deaths	More widespread water stress	Increased risk of food and water shortage and wild fires; increased risk of water- and food-borne diseases	Water shortages for settlements, industry and societies; reduced hydropower generation potential; potential for population migration
Number of intense tropical cyclones: increases	Damage to crops; uprooting of trees	Power outages cause disruption of public water supply	Increased risk of deaths, injuries, water- and food-borne diseases	Disruption by flood and high winds; withdrawal of risk coverage in vulnerable areas by private insurers
Incidence of extreme high sea level: increases	Salinization of irrigation and well water	Decreased freshwater availability due to saltwater intrusion	Increase in deaths by drowning in floods; increase in stress-related disease	Costs of coastal protection versus costs of land-use relocation (see also tropical cyclones above)

Source: Parry et al., 2007.

Fig. 1. Number of extreme weather disasters in the UNECE region and globally, 1980–2008

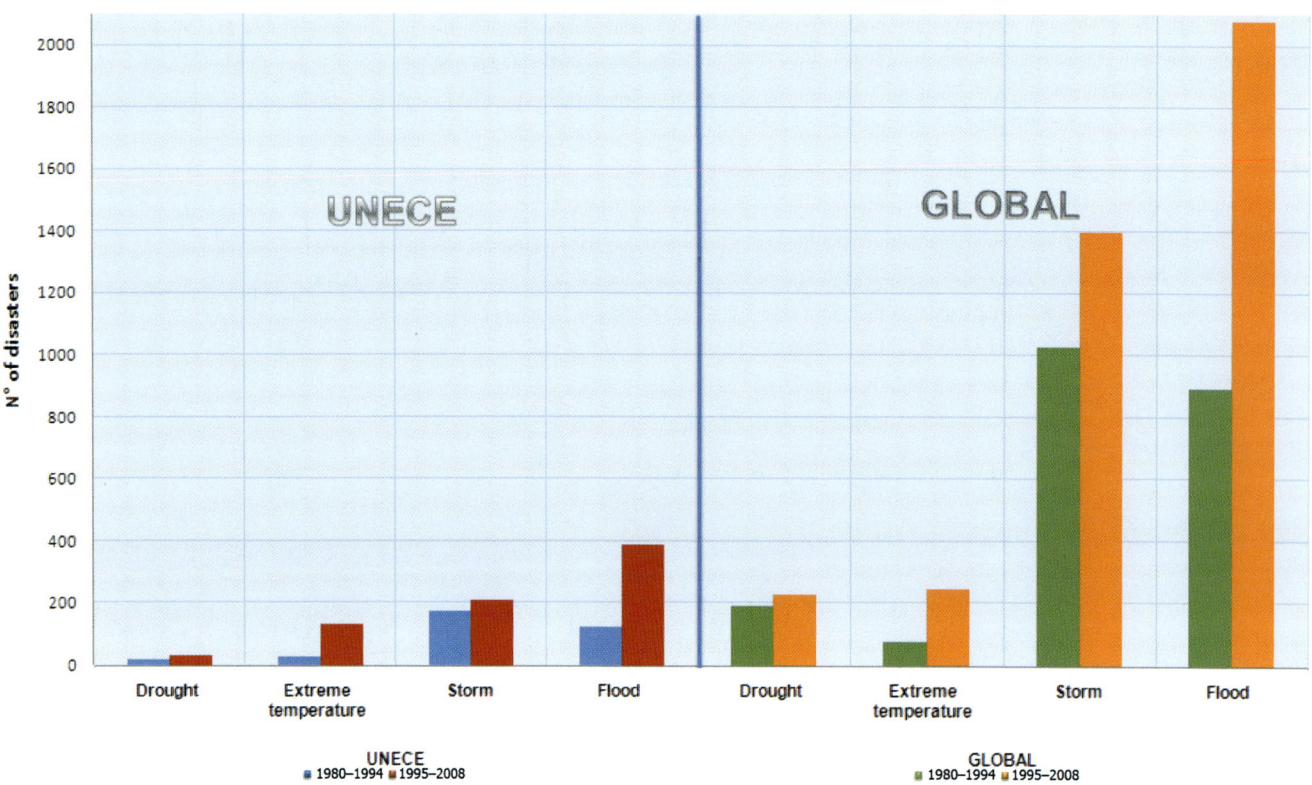

Source: adapted from EM-DAT (2009).
Note: UNECE region: European region plus United States and Canada.

Fig. 2. Number of people affected by extreme weather disasters in the UNECE region, 1970–2008

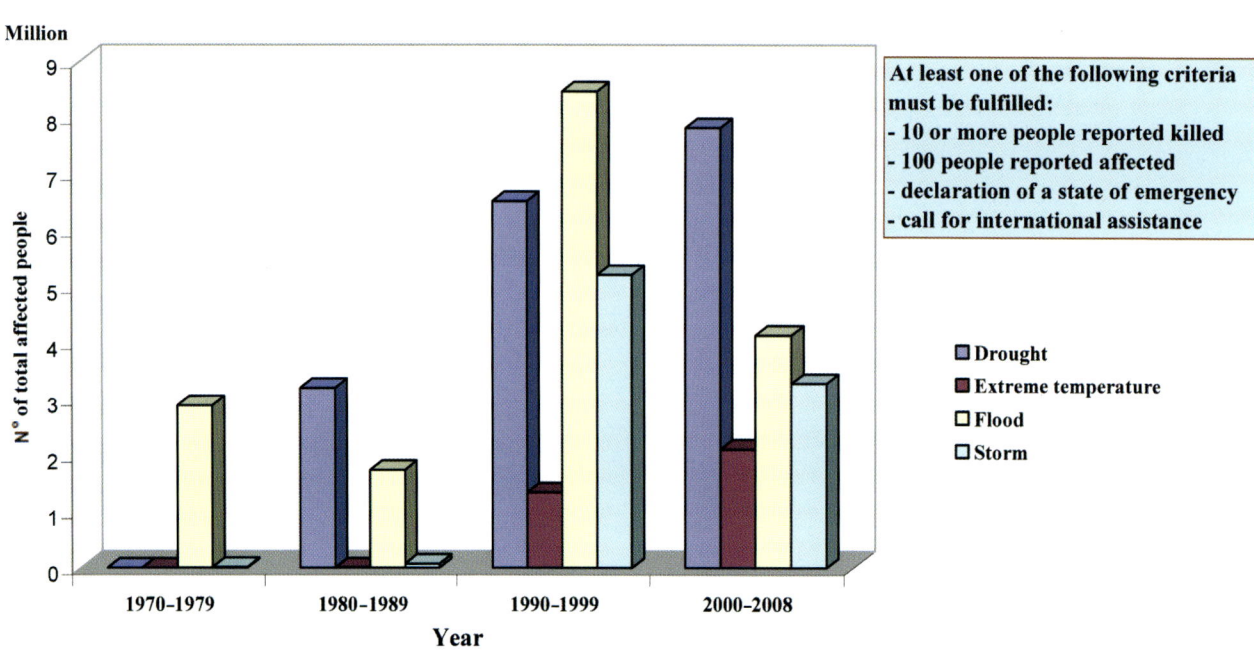

Source: adapted from EM-DAT (2009).
Note: UNECE region: European region plus United States and Canada.

impacts and inflict socioeconomic damage, including environmental clean-up and recovery costs. Disaster conditions are also liable to damage services and transport infrastructure, as well as houses, and to affect economic activities and lead to health system expenditure.

A European Environment Agency (EEA) report (EEA, 2008) also shows that the average number of annual disastrous weather- and climate-related events in Europe increased by about 65% between 1998 and 2007.

In terms of financial losses for central Europe, Table 2. (Central Europe's most costly flood catastrophes, 1993–2006) shows costly flood disasters in the period 1993–2006 and the related insured losses.

In Europe, overall losses caused by weather- and climate-related events increased during the period between 1980 and 2007 from an average in the previous decade (1980–1989) of less than €7.2 billion to about €13.7 billion (1998–2007).

All the countries of central Asia and the Caucasus are highly vulnerable to meteorological and hydrological drought. A severe and widespread drought in 2000–2001 wiped out at least between 10% and 26% of crop and livestock production in Armenia, Georgia and Tajikistan (3–6% of overall GDP) (World Bank, 2006).

Table 2. Central Europe's most costly flood catastrophes, 1993–2006

Period	Affected area	Losses * (original values, in € millions)	
		Overall losses	**Insured losses**
1993	France, Italy, Switzerland	1 245	415
	of which northeast Italy	520	
	Switzerland	350	200
1993	Rhine (Germany, Belgium, the Netherlands, Luxembourg, France)	1 765	705
	of which Germany	530	160
1994	Northern Italy	7 470	50
1995	Rhine (France, Germany, Belgium, Luxembourg, The Netherlands)	2 700	700
	of which Germany	270	100
1997	Oder (Poland, Czech Republic, Slovakia, Germany, Austria)	5 500	745
	of which Poland	3 205	410
	Czech Republic	2 020	165
	Germany	330	32
1999	Northern Alps and northern foothills of the Alps (Germany, Switzerland, Austria)	460	120
	of which Germany	340	70
	Switzerland	80	50
2000	Italy, Switzerland	10 000	560
	of which Italy	8 000	350
	Switzerland	390	210
2002	Elbe, Danube	16 825	3 465
	of which Germany	11 600	1 800
	Austria	2 445	410
	Czech Republic	2 445	1 225
2005	Austria, France, Germany, Hungary, Slovenia, Switzerland	2 685	1 430
	of which Switzerland	1 950	1 300
	Germany	175	40
	Austria	515	105
	Hungary	40	No details
	Slovenia	4	No details
2006	Elbe, Danube	390	40
	of which Germany	85	17
	Austria	21	3

© 2010 Münchener Rückversicherungs-Gesellschaft, Geo Risks Research, NatCatSERVICE
Source: 2010 Münchener Rückversicherungs-Gesellschaft, Geo Risks Research, NatCatSERVICE
Notes: * Original values, not adjusted for inflation; converted into € at month- and year-end exchange rates.

In the same period the hardest-hit communities of Armenia, Azerbaijan, Georgia, Tajikistan and Uzbekistan required food, drinking water, and agricultural relief supplies costing about US$ 190 million. The economic impact of agricultural losses and the subsequent relief operations is illustrated in Table 3 (Minimum losses from agricultural drought and cost of relief operations in central Asia and the Caucasus, 2000–2001).

Windstorm losses across 29 countries in Europe show an increase of more than 200% in the past 20 years compared to 1970–1989 (Barredo et al., 2009).

Table 4 (Selection of major windstorm catastrophes in central Europe, 1990–2007) shows some details of overall and insured losses from severe windstorms in central Europe.

1.4 Extremes Are Not Only Defined by Direct Damage

There is a lack of knowledge regarding medium- and long-term impacts on the environment of extreme weather events, as well as the effects of unhealthy environments.

Many data sources support the goal of a more appropriate assessment of how much effort is needed to address the role of contamination (and of water and wastewater services as a source of contamination) in safeguarding healthy water, as well as food consumption and production. Assessment should also address the issue of how such intense and recurrent episodes will impair the (costly) environment and health policies that are in place to prevent water-related diseases.

Since 1998, major river catchments of European countries have been hit by several flooding episodes. This is illustrated in Fig. 3 (River catchments affected by flooding, 1998–2005). In the same EEA assessment, the data showed a higher average rate of global SLR in the past 15 years of about 3.1 mm/year, with an increased risk of impacts on water services infrastructure, tide surges and saline intrusion, and with potential impacts on coastal ecosystems, wetlands and water availability for domestic, agricultural and drinking purposes. All of these represent high vulnerabilities for drinking-water availability, the performance of wastewater treatment and desalination plants, and the ability of ecosystems to assimilate waste and pollutants.

It is worth recalling that marine and inland aquatic ecosystems are interconnected. Some inland aquatic ecosystems are linked to the ocean ecosystems which they affect, for example through nutrient inflows that cause high productivity in many coastal fisheries, but also negatively affect them by means of pollutants carried by the water. In addition, a number of marine fishery resources (for instance, fish and shellfish) need inland water ecosystems, including estuaries and lagoons, to complete their life-cycles. The proliferation of harmful phytoplankton in marine ecosystems can cause massive fish kills, contaminate seafood with toxins, affect local and regional economies and upset the ecological balance.

Besides flooding, increasing water scarcity and droughts in many parts of the world may further limit access to water for sanitation, and consequently exacerbate health impacts and limit the ability of natural ecosystems to assimilate waste. In large cities, water scarcity is reducing the self-cleaning capacity of sewers and flooding is exacerbating stormwater overflows, resulting in pollution. Droughts or shortages of water can also affect bathing-water quality, because the decreased stream flows do not sufficiently dilute sewage and wastewater loads, causing an increase in pathogen numbers and untreated chemicals.

A WHO Regional Office for Europe report showed that diarrhoeal disease caused over 13 500 deaths in children aged under 14 years in the eastern European and central Asian countries of the European region in 2001 (WHO Regional Office for Europe, 2007), with a strong association with poor drinking-water quality and hygiene and a lack of sewerage and sanitation.

Table 3. Minimum losses from agricultural drought and cost of relief operations in central Asia and the Caucasus, 2000–2001

Country	Agricultural losses (million US$)	% of GDP	% of agricultural GDP	Relief operations (million US$)
Armenia	110–143	2.7	10.1	19.2
Azerbaijan	110	1.0	6.0	n.a.
Georgia	350–460	5.6	25.5	40.9
Tajikistan	100–159	4.8	16.8	104.4
Uzbekistan	130	0.8	2.4	22.9
Total	800	2.0	7.9	187.5

Source: World Bank, 2006.
Note: n.a.: not applicable.

Table 4. Selection of major windstorm catastrophes in central Europe, 1990–2007

Year	Period	Event	Affected area	Losses *(original values, in € millions)	
				Overall losses	Insured losses
1990	25–26/01/1990	Winter Storm Daria	Germany	1 000	690
	3–4/02/1990	Winter Storm Herta	Germany	500	250
	25–27/02/1990	Winter Storm Vivian	Germany	1 000	435
			Austria	100	66
			Switzerland	80	50
	28/02–01/03/1990	Winter Storm Wiebke	Germany	1 000	450
			Austria	100	66
			Switzerland	80	50
			Italy	20	No details
1992	21/07/1992	Severe weather event	Switzerland	76	40
	28/08/1992	Hail	Germany	90	72
1994	04/07/1994	Hail	Germany	425	325
		Winter Storm Lore	Germany	340	220
			Austria	5	3
			Switzerland	10	No details
1995	21–23/07/1995	Winter Storm Emily	Germany	400	300
1999	3–4/12/1999	Winter Storm Anatol	Germany	300	100
	26/12/1999	Winter Storm Lothar	Germany	1 600	650
			Switzerland	1 500	800
			Northern Italy	50	No details
2000	3–4/07/2000	Hail	Austria	100	90
2001	5–7/07/2001	Severe weather event, tornado	Czech Republic	17	6
	7–8/07/2001	Severe weather event, tornado	Northern Italy	200	35
2002	26–27/02/2002	Winter Storm Anna	Germany	580	350
	24/06/2002	Hail	Switzerland	220	170
	05/08/2002	Hail	Northern Italy	70	51
	26–30/10/2002	Winter Storms Jeanett, Irina	Germany	1 700	1 200
			Czech Republic	20	10
	16–17/11/2002	Severe weather event, windstorm	Austria	100	70
	15–28/11/2002	Severe weather event, landslides	Switzerland	170	50
2003	2–3/01/2003	Winter Storm Calvann, floods	Germany	280	90
	29–31/08/2003	Severe weather event, landslides	Northern Italy	455	5
2004	09/08/2004	Hail	Slovenia	15	No details
	20/11/2004	Winter storm	Slovakia	200	8
2005	7–9/01/2005	Winter Storm Erwin (Gudrun)	Germany	210	150
2006	16–17/08/2006	Hail, severe weather event	Austria	80	55
	28–29/06/2006	Hail, severe weather event	Germany	390	230
2007	18–20/01/2007	Winter Storm Kyrill	Germany	4 200	2 400
			Austria	310	200
			Czech Republic	150	100
	20–21/06/2007	Severe weather event	Switzerland	85	60
2008	1–2/03/2008	Winter Storm Emma, storm surge	Germany	750	400
			Austria	350	200
	28/05–02/06/2008	Severe Storm Hilal, hailstorms, flash floods	Germany	1 100	800
	23–24/07/2009	Severe storms, hailstorms	Switzerland	700	480
			Austria	350	220

© 2010 Münchener Rückversicherungs-Gesellschaft, Geo Risks Research, NatCatSERVICE
Source: 2010 Münchener Rückversicherungs-Gesellschaft, Geo Risks Research, NatCatSERVICE.
Notes: * Original values, not adjusted for inflation; converted into euros at month- and year-end exchange rates.

In the same period, eastern Europe, Caucasus and central Asia (EECCA) countries experienced a significant drought.

In 2009 in Italy an extraordinary bloom of the cyanobacterium *Planktothrix rubescens* occurred in the Occhito basin, a 13-km^2-wide artificial reservoir with a storage capacity of over 300 000 m^3 of water. Maximum algal density exceeded 150 million cells/litre and associated microcystin (MC) production occurred in raw water used for human consumption in surrounding municipalities (home to about 800 000 inhabitants) (Lucentini et al., 2009).

There is a growing awareness of the socioeconomic cost of drought, but methodologies to assess its environmental health impacts remain lacking. One approach is shown in Fig. 4 (Concept framework for the assessment of drought impacts).

In terms of vulnerability assessment, it is widely known that any climate-related changes will impact on water quality and availability. Examples of such changes include those listed here:

- **increase in lake and river surface water temperatures,** causing changes such as the movement of freshwater species northwards and to higher altitudes, alterations in life-cycle events (earlier blooms of phytoplankton and zooplankton), and the increase of harmful cyanobacteria in phytoplankton communities causing a rise in threats to the ecological status of lakes and increased risks for human health;

- **reduced water flows** from shrinking glaciers and longer and more frequent dry seasons; decreased summer precipitation, leading to a reduction of stored water in reservoirs fed by seasonal rivers; biennial precipitation variability and seasonal shifts in stream flow; reduction in inland groundwater levels; an increase in evapotranspiration as a result of higher air temperatures; the lengthening of the growing season; and increased irrigation water usage;

- **increased household water demand in the hot season, water scarcity and drought,** impairing raw water sources' reliability, as it is altered by changes in the quantity and quality of river flow and groundwater recharge,

- **heavy effects on drinking-water quality** as a consequence of the decrease in pollutants being diluted (resulting from increasing water temperatures, and water scarcity/flow); and increased water flows displacing and transporting different components from the soil to the water through fluvial erosion; and

- **unsuitability of water for drinking and agriculture purposes,** as a consequence of saline intrusion.

Many drivers not related to climate, as well as global changes, also exacerbate some of the vulnerabilities associated with extreme events, when they affect, for example, hydrological systems and ecosystems, as well as economic and social systems.

Land-use changes play a role in the rainfall–runoff relationship. Deforestation, urbanization and the reduction of wetlands reduce the available water storage capacity and increase the runoff coefficient, leading to growth in flood amplitude and reduction of the time-to-peak. Urbanization has exacerbated flood hazards by increasing the number of sealed areas and infrastructures. The urbanization trend is also leading to unplanned slums with basic, poor or non-existent water and wastewater services.

In some countries, market-oriented policies have sought to increase crop production through huge changes in river basins, resulting in impaired drinking-water supplies.

The increase of water pricing is leading poor people, especially in small communities and rural areas, to use old, unsafe wells and unsafe new sources, such as untreated recycled water.

1.5 Extremes and Water Supply Sanitation: Old Problems, New Risks and Challenges

The main international and European scientific organizations have pointed out the potential impacts of climate change and extreme weather events on water and wastewater services,[8] identifying vulnerable groups and vulnerable subregions. Evidence shows that water supply and sanitation utilities, including all infrastructure elements of water abstraction, catchment areas, reservoirs, treatment plants, drinking-water pipelines and distribution systems, as well as sewerage networks are key environmental determinants in these critical conditions.

Improper management of the infrastructure may also have negative impacts on general water management and may in turn affect healthy water and wastewater services.

Internal vulnerabilities of these services comprise a wide range of science and information, in terms of existing networks, quality of performance in critical conditions, implementation of technology development and safe delivery of the services. Furthermore, the potential impacts of the individual "extreme events" can vary, as a different individual process exists within each system category of the complex water and wastewater services systems.

A quantitative and qualitative analysis of water services in the European region still shows vulnerabilities resulting from older sanitation arrangements. At the EU level, more than 20 million citizens lack safe sanitation, and rural areas are still very vulnerable compared with urban populations, as illustrated in Fig. 5 (Percentage of the population with domestic connection to improved sanitation facilities in urban and rural areas, 2006).

---

8 In this book, the term "water services" is consistent with the definition given in Art 2 § 38 of the European Union Water Framework Directive (WFD): "… all services which provide, for households, public institutions or any economic activity: (a) abstraction, impoundment, storage, treatment and distribution of surface water or groundwater, (b) waste-water collection and treatment facilities which subsequently discharge into surface water".

Fig. 3. River catchments affected by flooding, 1998–2005

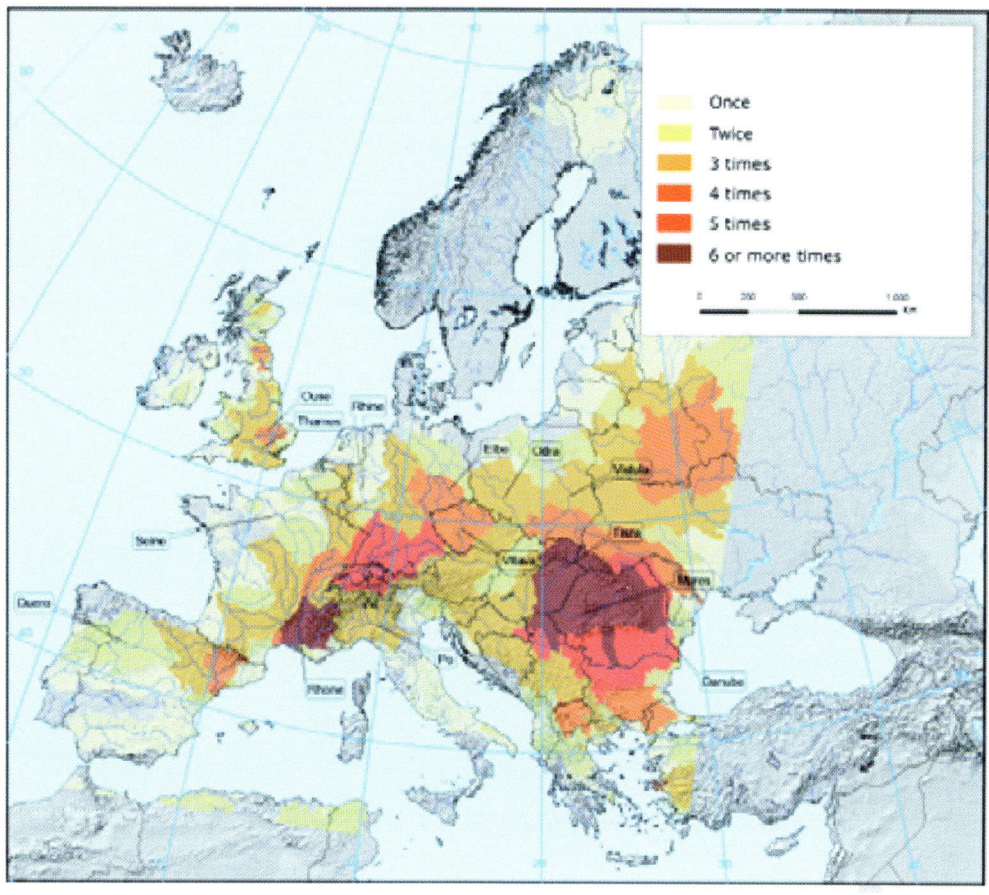

Source: EEA.

Fig. 4. Concept framework for the assessment of drought impacts

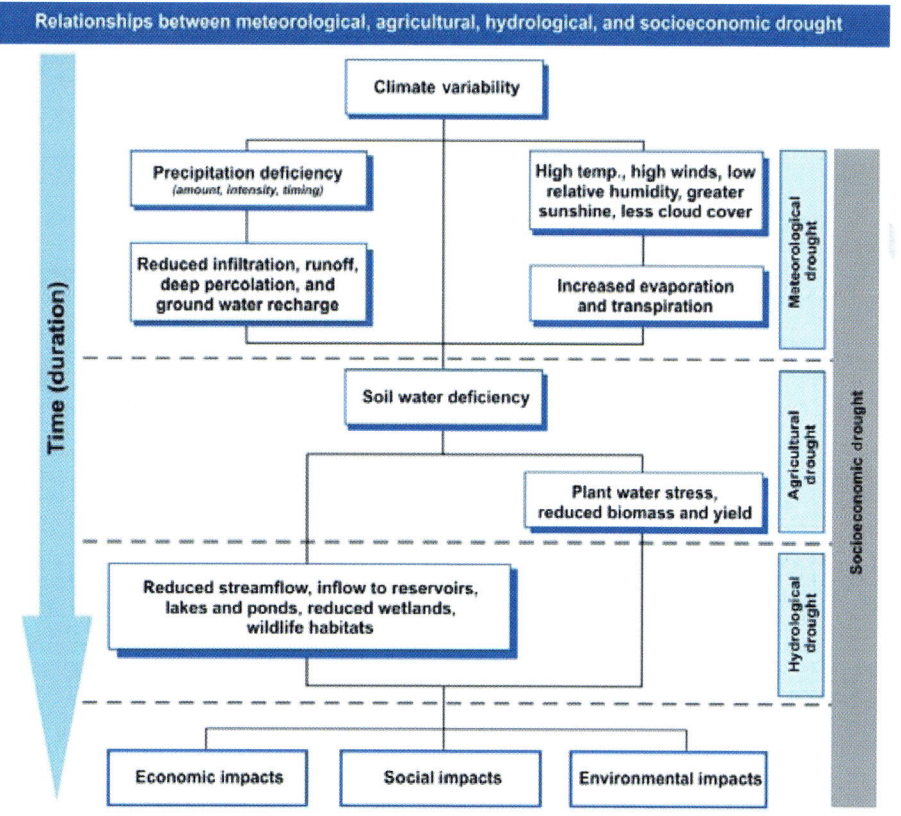

Source: NDMC.

Data up to 2005 show that in many countries of the European region the percentage of the population connected to wastewater treatment facilities ranged between 15% and less than 50%.

The EU UWWD (Council of the European Communities, 1991) dates back to 1991 and it is probably the most expensive Directive ever adopted, with investment costs of about €30 billion for the 12 newer Member States belonging to the EU since May 2004. However, not all Member States (EEA, 2005a; BIPE, 2006) are yet in compliance with all its provisions.

It was stressed that (WHO Regional Office for Europe, 2009):
Wastewater from households and industry places a significant pressure on the water environment through the release of organic matter, nutrients, hazardous substances and pathogenic microorganisms. The majority of the European population lives in urban agglomerations (three quarters in 1999); a significant proportion of urban wastewater is collected in sewers connected to public wastewater treatment plants. Contamination of aquatic resources by wastewater reduces the possible use of the recipient waters for a variety of applications: treatment of water to raise it to drinking-water standards may become technologically more challenging, while direct use in irrigation may pose specific health risks.

One of the main conclusions of the Conference of EECCA Ministers of Economy/Finance and Environment, "Financing water supply and sanitation in eastern Europe, Caucasus and central Asia" (Yerevan, Armenia, November 2005), (OECD, 2005) was:
In the countries of eastern Europe, Caucasus and central Asia (EECCA countries), problems of access to water services are rooted in the history of that region. Ambitious investment programmes led to the development of extensive networks of water infrastructure in urban and rural areas. However, these networks were often poorly designed and constructed, and they have not been adequately maintained. As a result, water supply and sanitation infrastructure has seriously deteriorated in most countries in the region and even collapsed in some places, with potentially calamitous consequences for human health, economic activity and the environment.

Improved sanitation or wastewater treatment will improve water availability in several ways. As improved treatment results in wastewater that is less polluted, the chance of polluting surface waters or shallow aquifers is reduced. These surface waters and shallow aquifers may therefore be used as a source for the production of drinking water or applied for other uses. Moreover, the ecological quality is less impaired. Wastewater from treatment plants may also be used directly, for example tertiary treatment provides wastewater that may be suitable for agricultural or industrial use. At the transboundary level,

Fig. 5. Percentage of the population with domestic connection to improved sanitation facilities in urban and rural areas, 2006

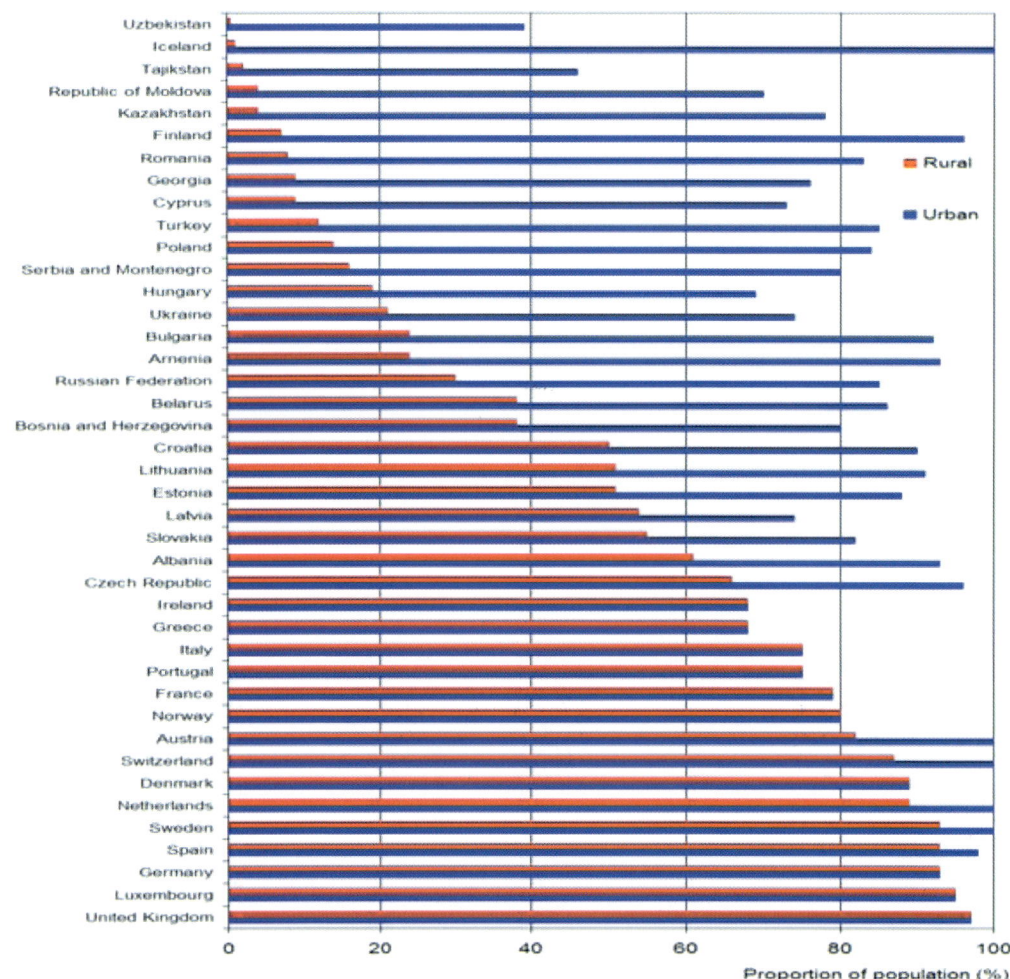

Source: WHO/UNICEF (2008).

improved wastewater treatment reduces conflict potential as upstream water uses have less impact on downstream use.

Poor design and maintenance of water and sanitation system infrastructure, on the other hand, may lead to serious pollution. Increased or intensified precipitation may, for instance, lead to system overflow, in which untreated wastewater reaches the surface waters of shallow aquifers. As a result, the use of these waters may have to be excluded from the infrastructure for several purposes. Prolonged periods of drought may lead to malfunctioning of the treatment systems and may result in pollution of surface waters or shallow aquifers. Leakage of pipelines and other distribution systems – through corrosion, mechanical stress or physical breaking, as a result of poor design or insufficient maintenance – results in loss of precious drinking water or pollution of shallow aquifers or surface waters. This in turn affects the availability of water of sufficient quality for use as drinking-water and for other purposes.

Conflicts exist between energy and water. Energy production uses water for fuel production processes, and for extraction and refining. Hydropower production alters the natural variability of surface waters and their consequent availability for other uses, such as agriculture and drinking-water. Finally, electrical power plants use water for cooling.

Water services also need energy. Extraction and transport of water, production of drinking-water (including desalinization), and delivery and return of water all require energy, as does the treatment of wastewater.

Energy production should therefore aim to develop closed systems in which process water is re-used, if necessary after treatment. This reduces the need for water, as well as the impacts of wastewater on surface waters and aquifers. Water services should aim for energy efficiency. Biological treatment, for instance, includes the possibility of producing biogas that in turn can be used for energy to support the treatment process.

Crucially, water pricing is another factor that is an important health determinant, to which all vulnerabilities mentioned thus far are directly related.

The costs of providing household, industrial and agricultural water services are increasingly covered through pricing mechanisms. Tariff structures for water supply and wastewater treatment have a role in increasing the cost–effectiveness of resource use. Water users such as the agriculture and industrial sectors can strike a balance between their use of water and the income that this will generate. For industry, it may lead to increased self-treatment and effluent re-use, the use of cleaner technologies and the reduction of waste generation. Households may become more self-aware in their use of water.

The cost of producing and delivering clean water to urban areas greatly depends on the proximity of raw water sources, the degree of purification needed and the settlement density of the area being served. The cost of providing sewerage and treating wastewater also depends on settlement density, as well as on the characteristics of the influent and the required quality of the effluent.

Water pricing may be achieved in different ways:

- as pollution charges for discharging effluent to natural waters; the charges can be based on volume only, on the effluent's pollution content, or on the cost of measures to prevent pollution of surface waters;

- as abstraction charges for ground or surface waters (or both); these are typically based on the maximum withdrawal rate permitted by an abstraction licence or on the actual volume withdrawn, but can also be based on the source (ground or surface), the availability of water in place or time (seasonal), or the type of user (i.e. agricultural or industrial users often benefit from exemptions);

- as service fees for domestic and industrial water services, to cover the operational and maintenance costs of operating water facilities; such charges can be volumetric or based on flat rates (for instance, linked to property values). Wastewater treatment fees are sometimes calculated as a fixed proportion of the water supply bill, or may vary with the volume of water actually supplied.

Costs for water services provision are likely to increase because of the need to meet existing and future drinking-water standards; refurbish or replace pipe networks, which are often inadequately maintained; upgrade sewage treatment standards; separate sewage from stormwater networks; and treat urban stormwater and wet-weather sewage overflows.

Social considerations must be included in water pricing, especially in less -developed countries. If water is sold at real cost this could represent a large fraction of household budgets and can reduce users' willingness to pay. As a consequence, users may turn to alternative (unsafe) sources of drinking-water, which may jeopardize their health, and they may seek means to dump wastewater, which may affect surface or other water bodies. In this scenario everything should be done to protect health, to overcome inequities and vulnerabilities and to foster the sustainability of natural resources and ecosystems.

Traditionally, water and wastewater services were built to protect people from unsafe water, and to protect the environment from dangerous pollution.

Under extreme conditions, even "gold-standard" technologies only barely manage to meet these goals. In addition, in practice their performance also has to cope with all the (by now familiar) climate-related factors, as well as those not related to climate.

The World Business Council on Sustainable Development (WBCSD) stated that (WBCSD, 2008) "[A]daptation will not reduce the frequency or magnitude with which climate change events occur, but will protect business and society against events such as drought, hurricanes and flooding".

Among the new challenges for water and waste services, the WBCSD has identified:

- greater demand as a result of increased temperatures and changes in supply;

- the need to cope with greater variability in river flow due to changes in temperature and precipitation, with possible damage to water supply infrastructure during heavy rains or droughts;

- salinization of coastal groundwater reservoirs;

- vulnerability of water services designed for steady conditions under new, highly variable conditions, such as floods and droughts;

- likelihood that water supply and treatment will become increasingly energy intensive and expensive, while climate change may cause conflicts between mitigation and adaptation policies.

Clearly identified risks also include: potential conflict between industrial water users situated in water-scarce areas over access to diminishing resources of decreasing quality; flooding of water supply works in riparian countries, leading to supply disruption; considerable costs of infrastructure upgrading; and associated damage and/or contamination.

The European Federation of National Associations of Drinking-water Suppliers and Waste Water Services (EUREAU), which includes both public and private suppliers, and collectively supplies water to around 405 million European citizens, also recognized (EUREAU, 2008):

> Water, as one of the most important enablers of economic and social development and public health [...] needs to be resilient to climate change and shifting weather patterns. There will be increasing pressure on water resources in the coming decades. Climate change represents a key challenge for the water sector in terms of availability of water, flooding in urban areas, and impacts on water and waste water treatment systems and assets.

The resilience of service infrastructure and of any technological adaptation measures – including early warning and monitoring systems – is also very important, but, within the framework of such complex scenarios of risks and drivers, cannot be simply left to engineering solutions and financial investment.

Science and technology development should also be accompanied by:

- active involvement of relevant stakeholders, such as utilities, land users and water resource managers;

- a facilitating mechanism for cooperation frameworks, tailored to local needs, with the aim of ensuring local capacities are furnished with the correct tools to perform, and enhancing providers' performance, knowledge and awareness of new environment and health risks;

- a commitment to multisectoral policy response, in order to cope with changes in adaptation strategies.

The health and policy relevance of this is underlined by the recent WHO headquarters Vision 2030 initiative (WHO, 2010) which clearly states:

> The ensuing adverse impacts on water and sanitation services constitute a clear and present danger for development and health.

New evidence, translated into new advocacy, is needed to raise the awareness in governments, international agencies, nongovernmental organizations and communities about the links between climate change and water and sanitation services, and the consequences for health and development. In a context of relative uncertainty associated with climate change projections, policy responses will have to be formulated based on our current knowledge to address these impacts and consequences.

Appropriate adaptation measures for utilities infrastructure form part of risk management strategies that would benefit from additional crucial assistance, such as adequate information tools for early warning and vulnerability assessment, mass communication strategies, and strengthening the coping capacities of environmental monitoring, early warning and disease surveillance.

All these issues are addressed in subsequent chapters.

1.6 CONCLUSIONS

Without any doubt, water supply and sanitation, together with energy, define the improvement of living conditions in human history.

There is also significant consensus on the importance of water supply and sanitation systems to environmental issues and health problems, social services, poverty alleviation, sustainable water resources management, food production and security, drinking-water supply and water-related natural disasters.

If weather and climate behave in extreme ways, water services systems (water supply, sewerage and wastewater treatment) will lose much of their environment and health benefit, becoming instead a significant source of chemical and biological contamination of ecosystems, water bodies and soil. Such contamination may sometimes be irreversible, and it may affect areas beyond local and national borders.

Abnormal weather, climate and hydrology can pose serious threats to sustainable livelihoods and to the health of exposed populations.

This is already happening in the European region, and has been for the past 20 years: evidence shows that floods, windstorms, droughts and extreme temperatures are already severely affecting the region, in line with the global trend.

Preventive actions need strengthening in order to limit direct damage and ensure safe basic survival needs are met, such as water, sanitation and medical assistance in emergencies, as well as to counteract the health hazards of extreme temperatures, water scarcity, chemical and biological contamination of water and food, and infectious diseases.

There is also urgent need for a centralized database to monitor the direct socioeconomic impacts of extremes in the European region and to help with the development of tools to assess the medium- and long-term impacts on the environment. There is also need to focus efforts on arguing the case for water and wastewater services to cope with both climate-related global and local drivers, as well as those not related to climate. Plenty of data exist to encourage a more appropriate assessment in order to address the role of contamination – and of water and wastewater services as sources of contamination themselves – in making possible healthy levels of water and food consumption and production. The data also encourage a new assessment of how intense and recurrent episodes of extreme weather will impair the costly environment and health policies introduced to prevent water-related diseases.

Early conclusions suggest that, in fostering the assessment of vulnerabilities, as well as risk management in extreme situations, many elements still need to be addressed, remembering that the performance of water supply and sanitation is the end point of pressures from extreme weather, climate-related drivers and those not related to climate.

Yet vast efforts are needed to bring water supply and sanitation infrastructure to a reasonable level of functioning in terms of extreme weather situations.

Several challenges indirectly affect water utilities management:

- increasing costs for ordinary and extraordinary maintenance of systems;

- increasing costs for technology development and tools to cope with recurrent adverse weather events;

- increasing costs for personnel training and early warning/modelling/forecasting;

- decision-making to solve conflicts regarding water users, for example large and small companies;

- new regulations on water supply and sanitation;

- communication issues.

All these issues need to be considered within a global context and should be managed within a general framework that includes transboundary cooperation, public authorities and agencies, all working towards integrated climate change adaptation strategies. A significant modification of strategies, infrastructure, systems and practices will be needed. Such approaches are currently being adopted by many water and wastewater utilities, but appropriate adaptation measures for utilities infrastructure and the overall system ability to cope in extreme weather event situations should be carefully assessed in view of emerging environmental health risks. Knowledge of these issues should be made available within water utilities companies in order to facilitate direct involvement in developing adaptation strategies.

Chapter 2

BASIC DISASTER PREPAREDNESS AND EARLY WARNING
Giacomo Teruggi
World Meteorological Organization

Aftermath of a flooding event, Tbilisi, Georgia, 2009
© National Centre for Disease Control and Public Health, Tbilisi (Georgia)/Nana Gabriadze

Basic Disaster Preparedness and Early Warning

2.1 Key Messages

Disaster preparedness and early warning are essential to limit the effects of extreme weather events. Key issues for the development of adequate preparedness are summarized below.

- The effectiveness of risk reduction in extreme conditions relies upon a commitment to apply integrated risk management principles in development planning, the existence of well-defined institutional responsibilities, a democratic process of consultation, and an information and awareness campaign. It moves beyond disaster response and reaction, towards risk anticipation and mitigation.

- Preparedness plays an essential role in the whole process, focusing on technology and trained staff needs in terms of procedures and tools necessary to coping effectively with a disaster.

- All information collected by monitoring networks should be made available to all responsible organizations including public health systems, the managers of reservoirs and dams, and water utilities operators who could be affected by the impacts, at both national and transboundary levels.

- Information development should provide adequate support to pre-event vulnerability assessment and post-event assessment of environmental and socioeconomic damage.

- Many monitoring, forecasting and management tools are already available for disaster preparedness planning.

2.2 Introduction

The United Nations International Strategy for Disaster Reduction (UNISDR) defines disaster risk management (UNISDR, 2009) as:

> The systematic process of using administrative directives, organizations, and operational skills and capacities to implement, strategies, policies and improved coping capacities in order to lessen the adverse impacts of hazards and the possibility of disasters.

This definition includes measures to avoid (prevention) or to limit (mitigation and preparedness) the adverse effects of hazards. Disaster management provides the means to prevent hazards becoming disasters. The process of disaster management involves a cycle of three phases: preparedness, response, and recovery. This is illustrated in Fig. 6 (Disaster management process).

Over recent decades there has been a trend towards developing information piecemeal to localize socioeconomic activities, high-density populated areas, sensitive sites (hospitals, nuclear power plants, industrial sites, etc.), and infrastructures in hazard-prone areas. This has increased the knowledge demand of risk assessment, promoting a "risk culture" which aims to assess, evaluate and reduce the escalation of risk linked to changes in land use and climate variability. International agreements have also sought to include disaster risk reduction strategy in the development process of adaptation strategies to climate change (for example, the United Nations HYOGO Framework for Action).

Risk in general, and in extreme weather conditions in particular, is the result of three factors: the magnitude of the hazard, the degree of exposure to the hazard, and overall socioeconomic and environmental vulnerability. This is illustrated in Fig. 7 (Components of risk).

Fig. 6. Disaster management process

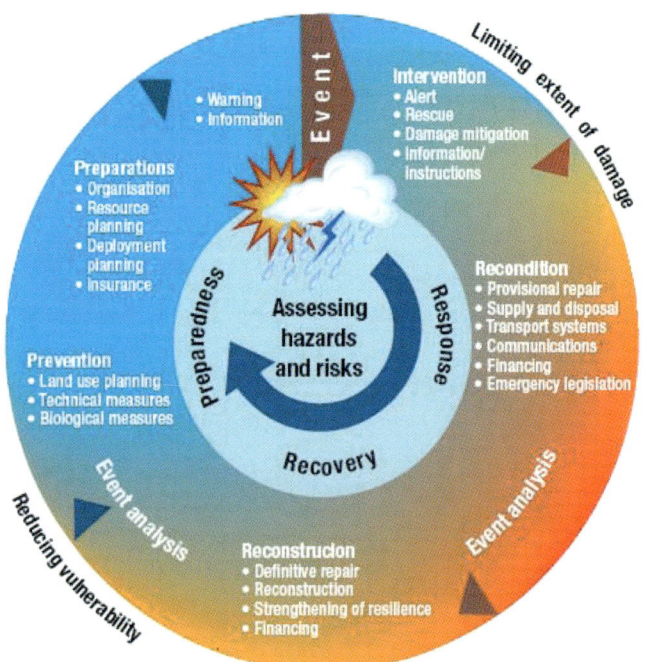

Source: Swiss Confederation National Platform for Natural Hazards (2001).

Fig. 7. Components of risk

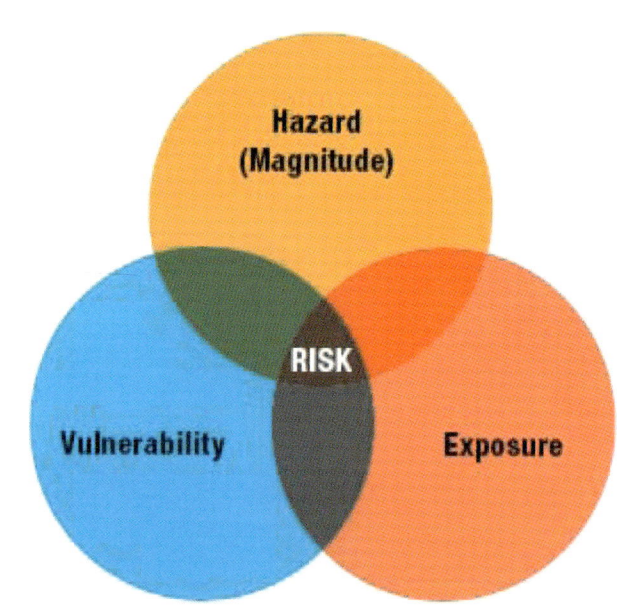

Source: WMO/GWP Associated Programme on Flood Management (http://www.apfm.info, accessed 15 March 2011).

Preparedness plays an essential role in the whole process, focusing on technology and trained staff needs in terms of the procedures and tools needed to cope properly with a disaster.

The role of water supply and sanitation utilities in extreme weather conditions is crucial, in preparing and implementing both a preparedness plan and adaptation strategies, since it holds real information and technical capability about the management of water catchments, adduction, storage, treatment, distribution and quality.

2.3 Information Needs: from Risk Assessment to Risk Reduction

The key principle for any development of preparedness and response plans, as for any effective warning system, is to focus on the three different and complex components of risk – hazard, exposure and vulnerability.

There are many tools for risk assessment already available at international and national levels. Some of them can present a sectoral approach, in terms of the type of extreme (e.g. flood plans, or heat waves) or of the impact (e.g. environmental, shelter safety, public health); others can provide a more comprehensive assessment approach.

2.3.1 Integration of information needs

Integration of available information is key to dealing with climate proofing. Combining the information listed below (geographical definition of exposed areas, population census, and survey of the economic and sensitive values and assets located in these areas like hospitals, industries, major infrastructure, nuclear power stations, etc.), it is possible to obtain an estimate of vulnerability of the elements at risk.

A more complete list of information needed for pre-vulnerability assessment is given in Table 5 (Pre-vulnerability assessment). In combination with data about frequency and magnitude of the hazard, different levels of economic losses can be calculated and expressed, for example in terms of damage per square metre per year ($/m^2$/yr) or in damage curves.

As an example, integrated flood risk maps help the final users to clearly identify the most endangered areas, infrastructures and utilities. Risk mapping summarizes and presents graphically the outcome of the risk assessment. Other than identifying the risk areas it allows users to have an overall and general view of all the components of risk in quantitative terms, identifying those that need to be addressed as a priority, so contributing to planning risk management measures and preparedness. An example of a risk map is shown in Fig. 8 (Risk map developed using geographic information system).

Communities and operators of water supply and sanitation utilities should take part in the participatory assessment of risks, vulnerabilities and capacities of supply, distribution and treatment systems linked to action planning by communities, linking them to local development plans.

Fig. 8. Risk map developed using geographic information system

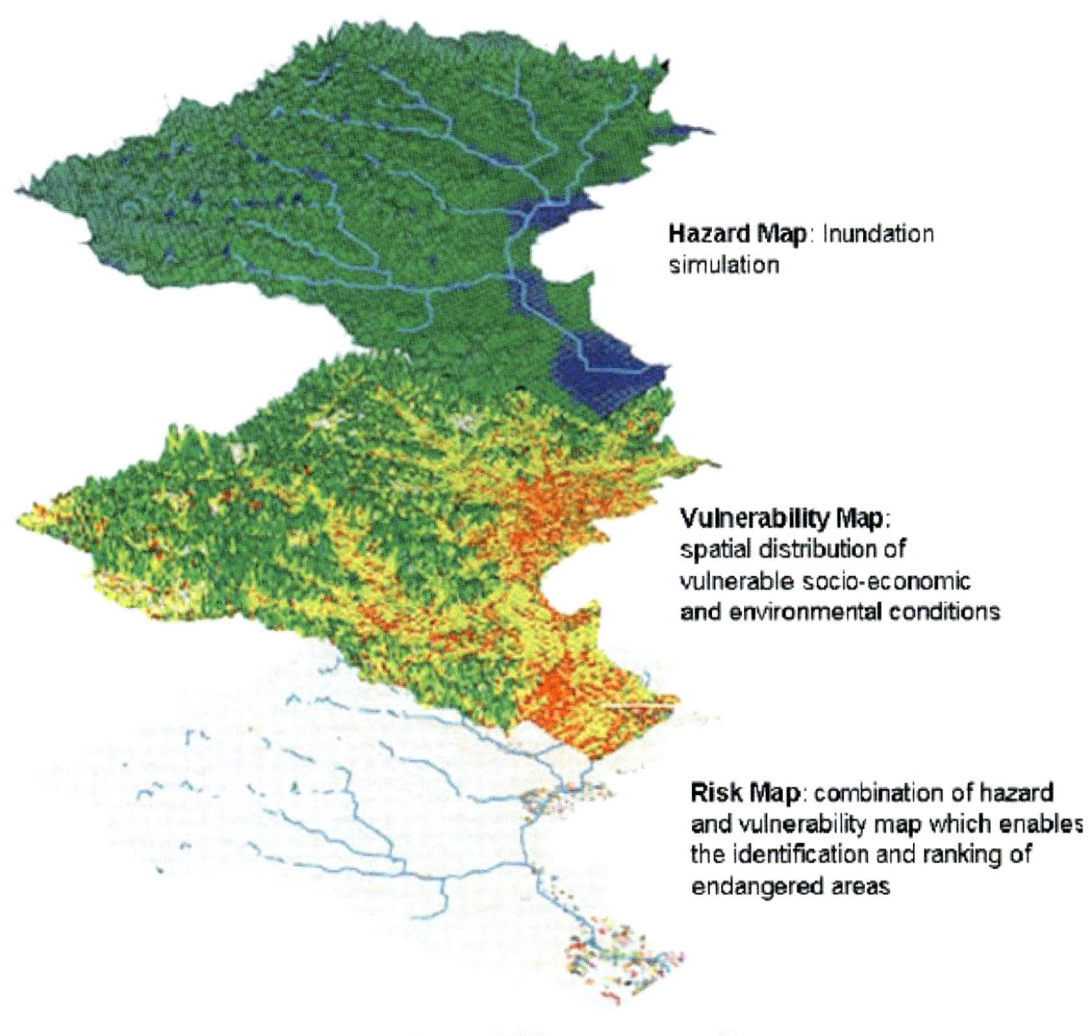

Source: adapted from Herath (2001).

Table 5. Pre-vulnerability assessment

Assessment	Tools
Hazard	**Hydrometeorological data** • Systematic streamflow measurements • Historical flood data • Annual peaks from rivers in the same region as the river for which hazard assessment is desired • Rainfall frequency data • Curves showing the largest observed floods as a function of drainage area • Temperature data • Evapotranspiration data • Soil moisture data • Groundwater data (including recharge) • Reservoir and lake levels • Seasonal weather forecasts **Policies** • Dam operation policy • Reservoir flood-control policy
Exposure	**Topographical data (WMO, 2006)** • Specialized geological data • Geomorphologic data • Soil studies • Floodable areas • Degree of urbanization • Drainage alteration • Aerial photographs • Satellite images **Census data** • Geographical distribution of population • Geographical distribution of specific categories of population (e.g. older people or known cardiopathic/ill people for heat waves) • Surveys • Density of housing • Siting of factories and infrastructures • Economic data
Vulnerability	As vulnerability is a combination of factors that can be physical, economic, social, political, environmental, technical, ideological, cultural, educational, ecological and institutional, these factors are often complex, dynamic and interrelated, mutually reinforcing/amplifying each other. See for example CEH Wallingford Climate Vulnerability Index described in Sullivan & Meigh(2005) and Sullivan & Huntingford (2009).

2.3.2 Post-event assessment of environmental and socioeconomic damage

Utilities managers are often pushed by urgent repair works needs and may miss such analysis. Cooperation with other experts/institutions working on the matter may further improve their performance for emergency response and relief coordination.

However, being ready to gather information right after the disaster is another key factor to improve prevention and preparedness. Damage analysis should start before clearing-up operations, while the traces of impacts are still visible. The results of lessons learnt will be useful for improving subsequent risk assessments, including those gauging the effectiveness of rescue operations, and the reconstruction phase.

Keeping this in mind, essential information for the emergency response at this stage would include:

- number of people affected by the malfunctioning of the utility due to the disaster;

- assets that have been damaged, restoration requirements;

- water quality data;

- assets at risk of being further damaged, based on the status of existing defences and, consequently, number of people at risk of being affected by further malfunctioning;

- status of lifelines (evacuation/access roads, electricity grid, fuel and disinfection product supply), hospitals and shelters;

- current and expected water levels at various locations, as well as weather conditions.

An assessment during the recovery phase is also necessary to understand administrative levels of responsibility for the response, (i.e. local or regional emergency response teams), or whether national or international assistance is required.

Further guidance on this issue is available from various sources, especially national and international bodies working on emergency response and relief coordination. A short list of readily available guidance material includes that listed here.

- *United Nations Disaster Assessment and Coordination (UNDAC) Field Handbook*, published by the United Nations Office for the Coordination of Humanitarian Affairs (UNOCHA), provides a rapid assessment methodology on a sectoral basis (UNOCHA, 2000).

- Office of the United Nations High Commissioner for Refugees *Handbook for emergencies* (UNHCR, 2007), provides checklists for initial assessments as well as guidance on the provision of safe drinking-water.

- *Community damage assessment and demand analysis* by the All India Disaster Mitigation Institute (AIDMI), provides guidance on a staged assessment process at the local level (AIDMI, 2005).

- *Post-disaster damage assessment and need analysis*, from the Asian Disaster Preparedness Centre (ADPC), provides ready-made templates for early reporting of damage and needs (ADPC, 2000).

2.3.3 Monitoring and forecasting

Besides hazard assessment, monitoring networks are also the basis of forecasting and early warning systems. Agencies/bodies with responsibilities for monitoring climate and water supply and sanitation usually perform the monitoring but the agencies responsible for collecting, analysing, and disseminating data and information may vary from one country to another. An analysis of the existing network and its task should be undertaken. Both in dry and wet periods, a reliable assessment of meteorological variables and of water availability and quality as well as the outlook for it in the short and long term depends on the data listed in Table 6 (Data needs for integrated assessment).

To be able to integrate in future a local network in the national one, adoption of WMO-recommended or NHMS-adopted standards is necessary. For specific information on how to establish an appropriate hydrometeorological network see WMO *Guide to hydrological practices* (WMO, 2009) or the WMO *Guide to meteorological instruments and methods of observation* (WMO, 2008).

All information collected by monitoring networks should be made available to all responsible organizations including public health systems, the managers of reservoirs and dams, and water utilities operators who could be affected by the impacts, at both national and transboundary levels.

For utilities managers, rather than duplicate networks, it is preferable to develop cooperative arrangements serving many purposes. In this respect, big water companies have their own monitoring systems usually linked to the company's remote control system and are technologically equipped to counteract power supply failure in extreme weather conditions. Smaller companies are usually not equipped with these devices/facilities, and connections among suppliers at local and transboundary levels should be properly arranged.

Tables and maps should be available providing details on monitoring locations, parameters, sensors, recorders, telemetry equipment and other related data. In addition, monitoring sites in adjacent basins should be inventoried. In low relief basins, data from those sites could be very useful. Analysis should be performed to identify subbasins that are hydrologically or meteorologically similar. It may also be the case that information from national monitoring networks is inadequate (architecture, technologies, etc.) for assessment at the local or transboundary level.

2.4 Tools for Disaster Preparedness Planning

2.4.1 Hydrological forecasting tools

Various hydrological tools are available for flood or drought hazard analysis. Hydrological analysis should be compulsory during the planning phase of water supply and sanitation utilities in changing climate conditions. The use of these hydrological tools depends mostly on the availability of adequate data and computation technology. It is however not the purpose of this publication to give a detailed explanation about these techniques, tools and data needs, but to offer a general overview and references. This is illustrated below in Table 7 (Hydrological forecasting tools).

Current hydrological forecast systems are quite affordable and powerful, but effectiveness relies on trained staff. These systems are capable of producing a broad range of forecasts, from stream conditions that will occur in a few hours to seasonal probabilistic

Table 6. Data needs for integrated assessment

Water quantity	Water quality	Meteorological variables
• Precipitation (rain gauges, low-cost weather radar) • Evapotranspiration • Soil moisture • Groundwater (piezometric hydrometers) • Streamflow • Reservoir and lake levels (level meters) • Snowpack	• Turbidity • Pathogen analysis • Chemical analysis • Saltwater intrusion in coastal areas* Water quality monitoring system should be capable of detecting sudden quality variations in different parts of the supply system (wells, springs, water intakes)	• Temperature • Wind speed and direction (anemometers) • Seasonal weather forecasts (climate outlook)

Notes: Monitoring should not restrict itself to the hydrometeorological factors, but also include potential impacts, e.g. areas subject to landslides or mudflows, glacier and snow melt and the resulting impact on reservoirs.

* Information required for such an assessment would include geology and hydrogeology maps, hydrology and catchment topography data, monitoring data, intake screen depth and stratigraphy, groundwater level variation and groundwater quality, with spatial references for all data.

outlooks targeted months in advance for larger rivers. Model system selection depends on the amount of data available, complexity of hydrological processes to be modelled, accuracy and reliability required, lead-time required, type and frequency of floods that occur, and user requirements.

With reference to forecasting tools for heat waves, they consist of calibrated models for the definition of meteorological maps; these models are usually fed by a series of parameters measured at ground level such as atmospheric pressure, humidity, wind speed, temperature. Through these maps pressure fields are simulated and/or forecast for definition, through mathematical algorithms, of rain, wind and temperature fields.

2.4.2 Early-warning systems

A warning means that the hazard is now a reality and that action has to be taken. Early-warning is vital for many response activities. The effectiveness of the warning depends on its reliability, the skilled interpretation of the warning signal, exchange and interactive cooperation between different early-warning systems (public health, meteorological, environmental, water managers and suppliers) and the ensuing emergency/rescue operations. The longer the lead-time, the more useful the warning, since the number of options for reaction is larger. The four elements of a people-centred early-warning system are shown in Fig. 9 (Four elements of people-centred early-warning systems).

To improve cooperation and avoid conflicts, an open and transparent communication mechanism between the warning manager, the disseminator, the receiver and down to the operators who should take action is a prerequisite. Relevant data and information on hydrometeorological variability and trends, water quality availability and health risks should be made available to water supply and sanitation utilities operators.

The main elements of the early warning chain are:

- detecting and forecasting impending extreme events to formulate warnings on the basis of scientific knowledge and monitoring, and consideration of factors that affect disaster severity and frequency;

- disseminating warning information, augmented by information on the possible impacts on people and infrastructure (i.e. vulnerability assessment), to the political authorities for further communication to the threatened population, including appropriate recommendations for urgent action; and

- responding to warnings, by the operators of the utilities, the population at risk and the local authorities, based on a proper understanding of the information, and subsequent implementation of protective measures.

Communication throughout the early warning chain must be two-way and interactive. Originators, disseminators and end-users must be in continuing contact with one another in order to make the system responsive to people's needs, priorities and decisions. The system has to adjust to users; not the other way around.

2.4.3 Management tools

2.4.3.1 Flood management tools

In the framework of Integrated Flood Management (IFM), the Associated Programme on Flood Management (APFM) is making an effort to provide guidance tools for flood managers and various other specialists working on the subject. A series of tools has been developed and made freely available for download at the APFM web site (http://www.apfm.info/ifm_tools.htm).

These tools are intended to help gain quick access to relevant technical guidance over the Internet. The guidance contained in these tools is intended to clarify the role and context of IFM in applying specific tools. The tools seek to incorporate various relevant materials previously scattered over the Internet and other sources.

Table 7. Hydrological forecasting tools

Frequency analysis	Flood or drought frequency analysis is used to estimate the relation between flood magnitude peak or minimal flow and frequency (WMO, 1989).
Regionalization techniques	To be used for frequency analysis when observed data are available only for short periods or few stations (regional frequency analysis produces results that are more reliable than frequency analysis at a single site – Potter, 1987).
Rainfall/runoff models	If stream flow data is limited, but on the other hand rainfall data is available, another tool that can be used for flood hazard analysis is rainfall/runoff models.
Hydrological modelling	In order to produce a flood forecast for the communities and locations at risk, there must be a hydrological modelling capability that uses meteorological and hydrological data. Hydrological models use real-time precipitation and stream flow data. The models translate observed conditions into future stream conditions.
Global climate models (and downscaling)	A combination of statistical techniques and regional modelling can be used to downscale climate models and to simulate weather extremes and variability in future climates.

Fig. 9. Four elements of people-centred early-warning systems

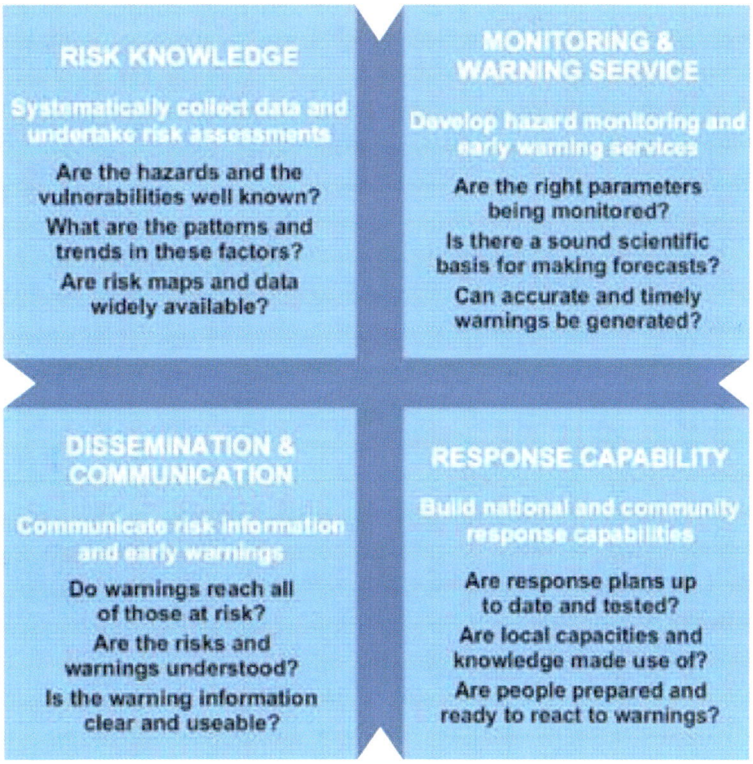

RISK KNOWLEDGE
Systematically collect data and undertake risk assessments
- Are the hazards and the vulnerabilities well known?
- What are the patterns and trends in these factors?
- Are risk maps and data widely available?

MONITORING & WARNING SERVICE
Develop hazard monitoring and early warning services
- Are the right parameters being monitored?
- Is there a sound scientific basis for making forecasts?
- Can accurate and timely warnings be generated?

DISSEMINATION & COMMUNICATION
Communicate risk information and early warnings
- Do warnings reach all of those at risk?
- Are the risks and warnings understood?
- Is the warning information clear and useable?

RESPONSE CAPABILITY
Build national and community response capabilities
- Are response plans up to date and tested?
- Are local capacities and knowledge made use of?
- Are people prepared and ready to react to warnings?

Source: UNISDR (2010).

2.4.3.2 Regional climate outlook forums

A regional climate outlook forum (RCOF) is a process, pioneered in Africa but still new to Europe, which brings together climate experts, sectoral users and policy-makers, to produce regional climate outlooks based on input from national hydro/meteorological services (NHMS), regional institutions, regional climate centres (RCCs) and global producers of climate predictions.

RCOFs assess the likely implications of the future climate (particularly droughts, heat waves, etc.) for the most pertinent socioeconomic sectors in the given region. RCOFs were originally designed to focus on seasonal prediction, and have significantly contributed to adaptation to climate variability and extreme weather events. The concept can be extended to develop capacities to adapt to climate change, and therefore to its consequences in terms of extreme weather events.

Forums include components ranging from scientific meetings of regional and international climate experts to developing a consensus for the regional climate outlook, typically in a probabilistic form. But probably more relevant for utilities managers is the fact that they involve as well both climate scientists and representatives from the user sectors (agriculture and food security, water resources, energy production and distribution, public health, and other sectors such as tourism, transportation, urban planning, etc.) to identify impacts and implications and formulate response strategies.

RCOFs also review obstacles to the use of climate information, experiences and successful lessons regarding applications of past RCOF products, and enhance sector-specific applications. The development of RCOFs requires good seasonal forecasting skills. These RCOFs then lead to national forums to develop detailed national-scale climate outlooks and risk information, including warnings which can be communicated to decision-makers and the public.

2.4.3.3 Involvement of utility managers in land-use plans

Knowledge of the hazard is a prerequisite for successful impact mitigation. Hazard and risk maps must be established, even if only a residual risk exists. Avoiding exposure to the hazard by keeping hazard zones free of intensive economic use is frequently recommended.

However, water sources and fertile soils encourage intensive human activities (agriculture, industry, and tourist areas) and settlement development, and skilled planning is required in hazard zones. Appropriate building codes and zoning restrictions should be established with the objective, if not of avoiding risk entirely, then at least of minimizing it in these areas.

With a multihazard approach in mind, it is also interesting to take into account the positive side-effects of some land uses (e.g. the conservation of pervious surface areas as farmland will also be effective as evacuation areas for other hazards). The prescription of what constitutes good practice depends very much on the type of hazard.

Safe delivery of water supply and sanitation services in critical conditions should be included in land-use plans and revised regularly (e.g. changes in land-use plans due to socioeconomic development together with utilities managers), taking into account the improved knowledge and technology to cope with hazard.

2.4.4 Hazard proofing

There are several available structural measures which must be adapted case-to-case to the type of hazard. Since total protection is not feasible, a predefined protection target or design standard has to be set. This target varies according to the economic and social values to be protected and according to the economic capacity of the society to protect them. These protection targets often are or become insufficient, either because knowledge of the hazard has improved (e.g. climate change) or due to an increase of the values to be protected (i.e. population development). In all cases prevention measures have to be accompanied by worst-case scenario emergency planning, which forms a key element of preparedness, including regular inspection and maintenance. More details on these measures will be given in the last two chapters of the *Guidance*.

Safe buildings are a key element in reducing vulnerability. Adequate building codes can improve resilience to several risks, including earthquakes, floods, landslides and tornadoes.

Stepping up investments in structural measures is necessary to reach "water security", that is, coping with too much, too dirty or too little water. Keeping water supply and sanitation utilities operational during extreme events will contribute to making the community more resilient to hazards. To this end, building in planned redundancy (e.g. building two access/evacuation routes rather than one, having back-up power generation capacity, a groundwater reservoir, etc.) should be promoted.

The involvement of managers of water supply and sanitation utilities is essential, since they should be able to operate continuously while responding to hazards. Therefore they need to be empowered; their management capabilities need to be strengthened; and their participation should be incorporated into disaster mitigation strategies.

In case of emergency, vital facilities, equipment and communications have to be replaced/repaired as soon as possible, even if only temporarily. In the first instance this includes the "lifelines" such as water supply, electricity supply, roads and telecommunications, hospitals and sewage systems. In the absence or temporary unavailability of safe water sources and sanitation facilities, utilities should be ready with alternatives.

It is also important to restore the water supply system if it suffers pollution from flooding events: a chlorine dispenser for disinfecting polluted water to an adequate level should be installed in critical sections of the system.

Other recovery measures may include mobile disinfecting plants or spare pumping stations installed in the same aquifer, exploiting water from a confined (unpolluted) artesian stratum.

2.5 Role of the Health System in Disaster Preparedness and Early Warning

Development infrastructure that forms the economic and social lifeline of a society, such as communication links, hospitals, and so on should be designed to cope with the most severe natural hazards and should work even in a disaster. Nor should this infrastructure increase the magnitude of the hazards.

Health systems need therefore to quickly recover their capacity to meet demand for service delivery in extreme weather conditions.

Extreme weather events affect health systems' operations and efficiency in different ways. Health facilities' built-in hazard-prone areas can be damaged, as can the access to them. The increased demand for health care could exceed the local public health capacity (including drugs, stockpiled vaccines, and trained personnel). Spontaneous or organized migration away from the area affected by the extreme events could shift the problem of exceeding the capacity of the health system to other areas. This also increases the potential risk of a critical outbreak of communicable disease, as well as increasing the risk of psychological diseases among the affected population.

Extended periods of drought or heat waves could lead to weakened resistance to various diseases.

The interruption of a health facility's operations after a disaster may be short term (hours or days), or long term (months or years). It all depends on the magnitude of the event and its effects on the health sector. The magnitude of an event cannot be controlled; its consequences, however, can be.

When planning a future health facility, the effects of these phenomena can be controlled if site selection is guided by sound information and criteria, and the design, construction, and maintenance can withstand local hazards.

During the 126th session of the Pan American Health Organization (PAHO)'s Executive Committee, it was decided to reduce the impact of emergencies and disasters on health through the following actions (PAHO, 2000):

- planning and executing public health policies and activities covering prevention, mitigation, preparedness, response, and early rehabilitation;

- providing an integrated focus addressing the causes and consequences of all possible emergencies or disasters that can affect a country;

- encouraging the participation of the entire health system, as well as the broadest possible intersectoral and inter-institutional cooperation, in reducing the impact of emergencies and disasters; and

- promoting intersectoral and international cooperation in finding solutions to the health problems caused by emergencies and disasters.

Table 8. Health system planning for flood preparedness

Type of activities	Health outcome and preventive measures
Pre-flood activities	• Long-term risk management: flood health prevention as part of multipurpose planning • Inter-institutional coordination • Infrastructure flood-proofing • Service planning risk zoning, risk mapping of health care and social care facilities, availability of communication and transport possibilities, emergency medical service preparedness, water and food supply planning for emergencies, evacuation organization, etc. • Awareness-raising campaigns targeting different groups in areas at risk • Capacity-building and personnel training for emergencies
Health protection during floods	• Prevention and treatment of infectious diseases, respiratory problems, injuries, mental health problems and skin and eye diseases – review and prioritize • Possible extra vaccinations for the general population • Communication campaign such as distribution of "boil water" notices, general hygiene advice and information on preventing mould, rodents, snake bites, and electrocution • Outbreak investigation where appropriate • Enhanced epidemiological surveillance of infectious diseases • Risk assessment of major environmental sources of contamination of health relevance • Intensify monitoring of drinking-water quality (tap) • Water and food provision
Long-term health protection	• Treatment for mould and other pathogenic exposures • Post-flood counselling (for anxiety and depression, for example) • Medical assistance • Enhanced cause-related surveillance • Research for future preparedness and response

Sources: Meusel et al., 2004; WHO Regional Office for Europe (2005).

As an example, a comprehensive reference for challenges that have to be met by the public health sector in floods is summarized in Table 8 (Health system planning for flood preparedness).

Communication and information to the public are also a key role of the health sector to prevent exposure to hazards in extremes (heat waves, cold spells, floods). For more details please refer to the next chapter.

- All key actors such as climate, environment, IT and health professionals should be helped to cooperate to assess and cope with vulnerabilities. In this regard, utilities managers should be involved in information sharing and preparedness activities, as they are themselves responsible stakeholders in coping with water and sanitation risks along with increasing demographic pressure, land use, overexploitation of resources, and climate change.

2.6 Conclusion

- The effectiveness of risk reduction in extreme conditions relies on a commitment to apply integrated risk management principles in development planning; the existence of well-defined institutional responsibilities; a democratic process of consultation and information; and an awareness campaign. It moves beyond disaster response and reaction, towards risk anticipation and mitigation.

CHAPTER 3

COMMUNICATION IN EXTREME WEATHER EVENTS
Benedetta Dell'Anno
Ministry for the Environment, Land and Sea, Italy

Handling media skilfully is essential
© WHO

Communication in Extreme Weather Events

·3.1 Key Messages

Experience shows that crisis management plans often focus on the technical aspects of resolving system failures, and limit communication to advisory warning on the use of individual measures to protect health. Yet in a crisis situation, people in a stricken area want to look, see and feel that effective leadership is being applied to resolve the problem and prevent its recurrence. Effective communication is a critical resource, especially during an emergency. Yet few managers of water supply and sanitation services have received any training in effective communication.

The following key messages may be helpful.

- The communication strategy, based on a multidisciplinary approach, should be part of the risk disaster management and adaptation plans for extreme weather events in order to share knowledge among different actors.
- Specific communication activities should be planned (before, during and after the event) and targeted at different groups at risk (e.g. the elderly, children, rural communities).
- Public authorities must be mainly responsible for elaborating and delivering the messages.
- The media are a key partner in communication.
- Communication should be a long-lasting and institutional process and not only a contingency tool.

3.2 Introduction: Importance of a Communication Strategy

People not aware of risks can slow down the emergency operations. Persuading people to evacuate in advance of a predicted threat is very difficult because they tend to discount warnings and it can exacerbate the problem. Appropriate information distribution and sound decision-making before and during weather emergencies are critical to saving lives, reducing injuries and protecting property.

Based on a multidisciplinary approach, a communication strategy can improve the effectiveness of the interventions and so the authorities (water managers) should consider including this strategy in the risk disaster management and adaptation plans designed to cope with extreme weather events. Different types of risk communication strategies should be developed, depending on the type of events (heat waves, cold spells, storms, flooding, etc.).

In order to improve the effectiveness of the measures, local authorities play a key role in undertaking work on preparedness and response to extreme events.

The communication strategy should include a plan of specific activities which should start **before** the crisis (pre-events activities) such as specific education programmes at schools, capacity building projects, training of personnel (media staff included) and awareness raising campaigns targeting different groups at risk and focused on vulnerable groups (the elderly, children, etc.).

During the event the communication campaign should provide the public with a unique, early and accurate announcement to foster trust in the institutions, to build public flexibility and to guide appropriate public participation to support the rapid control of the crisis. **After** the crisis the lessons learned should be considered in future planning as a useful tool for updating the communication strategy.

3.3 Communication Activities

Extreme weather events and their impacts on health and environment emphasize the need to keep the community and the general population informed about the associated risks for individuals and the protection measures being taken before, during, and after the event. The first steps are the collection and evaluation of the data. Then the authorities should provide information to the population on how best to protect themselves from extreme weather and climate events in general.

The public authorities are the main source responsible for elaborating and delivering the messages and it is advisable to establish a coordinating agency with a leading authority supported by a "risk communication team" consisting of the different actors involved such as water managers, media, nongovernmental organizations and other relevant stakeholders.

Awareness-raising campaigns on water and sanitation management and health outcomes should be addressed before the crisis happens. **Education** programmes in schools also play a key role. **Training** for the identified team should be focused on how to develop timely and effective communication skills to inform the public, partners, and stakeholders about recommendations. During public health emergencies the spokesperson's image and voice should be familiar to audiences to invite trust. He/she should be trained in how to build up and deliver the messages and should be informed about water and sanitation utilities and public health threats.

Along with increasing awareness and training, the **attitude change of the general public** should be promoted. Communication is a dialogue and requires close interaction between information providers and those who need the information. Beliefs and attitudes can be changed through openly pointing out and explaining the issue. After improving people's understanding, the receptiveness of climate change messages and creating ownership of the problem, behavioural changes can be promoted. Individuals and groups can be persuaded that they can make a difference in terms of their own lifestyle choices and in mobilizing their communities to reduce their risk of climate-sensitive diseases. In some countries certain groups of the population are more difficult to reach than others, for example the elderly, children, immigrants, rural communities and other groups. They need specific attention as well as a specific approach.

Appropriate risk communication procedures foster the trust and confidence that are vital in a crisis. The information should be provided early in a crisis and it should be transparent about what is known and unknown. That helps build **trust and credibility**. The first official announcement will establish the trust and this will guarantee the public acceptance of official guidance and trust in **institutions** and their recommendations.

The announcement should be **accurate, timely, unique, frank and comprehensive**. There is always a complex relationship between providing accurate information and providing it quickly. To wait for complete and verified information before releasing it to the public can generate an information vacuum which can give rise to speculation. Against that, to release information which has not been double-checked or is actually inaccurate runs the risk of misleading the public and undermining the credibility of a spokesperson. Understanding people is also critical to communicating effectively. To change people's beliefs and their **perceptions** it is necessary to understand what they think and why they act in the way they do. Communicators should develop proper messages to target the public, including information on how they can make themselves safer in a crisis.

Communication with the public and the news media should be planned, taking into account what information it is crucial to pass on in initial messages; what kind of messages should be delivered before, during, and after the event; what are the obstacles to effective communications and how to minimize them for specific audiences. Isolated communities should be identified and special tools developed to reach them.

3.4 Partnership in Communication

An effective risk communication strategy includes planning, preparation, messaging and working with the media, and the ability to manage the flow of information at each stage.

The **media** play a major role and are the most common channel of communication to the public. The best way to address the challenge is to establish regular briefings with the media and a trained key staff member to deliver explain and update all information.

Necessary information should be identified and appropriate leaflets, fact sheets and information materials should be developed. Dissemination of information can be achieved by means of the mass media (radio, television, and newspapers), web sites and short message systems via mobile telephones. The general public mostly accepts television advice messages and TV remains the main channel for reaching different population groups.

The **media communication strategy** should be part of the risk communication strategy and it should be planned in advance with the participation of the authorities, nongovernmental organizations and other relevant stakeholders. Cooperation with partners as an ongoing process is important if you want the public to take action. In order to deliver targeted and effective messages it is necessary to find out the reporter's needs.

Knowledge and media warnings alone are often not enough to persuade people of the seriousness of a situation. Interpersonal communication is very powerful, so the communicator should take into account **social networks** and develop a long-lasting cooperation with them.

3.5 Monitoring and Evaluation of the Outcomes

Before and during the events, the communication strategies and their tools should be tested practically and monitored, while an evaluation of the outcomes should be carried out after the event.

The ex post evaluation is crucial to measure the value and effectiveness of the activities, not least in terms of their cost. It is important to know what changes have happened as a result of communications: the level of awareness of the target audience; actions before/after extreme weather events, and gaps in communications that can be improved. Also, evaluation results are an information source for risk managers, decision-makers and the public. It is imperative that, after events, lessons are learned, including what did not go well and should be improved.

3.6 Conclusions

Appropriate communication and information to the public can reduce injuries and save lives.

Communication is a key issue to improve the prevention of harm to health as well as the effectiveness of interventions.

Chapter 4

VULNERABILITY OF COASTAL AREAS AND BATHING WATERS IN EXTREME WEATHER EVENTS

Gyula Dura
National Institute of Environmental Health, Hungary

Bathers forced out of Lake Balaton, Hungary, by a sudden summer storm
© National Institute of Environmental Health, Budapest (Hungary)/Gyula Dura

Vulnerability of Coastal Areas and Bathing Waters in Extreme Weather Events

4.1 Key Messages

Variations in the frequency and intensity of extreme weather events will pose serious challenges to the management of unique coastal ecological and cultural systems. Vulnerable coastal systems include fisheries, agriculture, tourism, marine and freshwater resources, health infrastructure, and municipal water supply and sanitation systems.

Both major extremes of global climate change can have serious impacts on coastal areas, as detailed here.

- Drying and water scarcity may result in the over-exploitation of groundwater resources, reducing their availability as well as impairing their quality (through contaminant concentrations) with harmful consequences for water supply to the population, agriculture and energy production.

- Extreme rainfall and storms may result in increased runoff, river discharge, more intense erosion and the mobilization of chemical and biological contaminants by surface runoff from urban and agricultural areas.

- A combination of rising sea level and more intensive coastal storms would create the highest environmental and health risk stemming from a salinization of water supplies, including aquifers used for drinking-water. The major problem is that, in most if not all of the coastal regions, groundwater is a key source of water supply, especially of drinking-water (more than 2 billion people worldwide depend on groundwater).

- Saline water intrusion, accelerated by both the rising sea level and the over-exploitation of groundwater resources in a drying climate, poses both quantitative and qualitative risks to the population. Extreme storm surges combined with rising sea level could result in much higher rates of coastal erosion, which would in turn further increase saline water intrusion. A 5% increase in salt content will rule out many important uses, including drinking-water supply and the irrigation of crops, parks and gardens, and will threaten groundwater-dependent ecosystems.

One approach to better understanding the vulnerability of coastal areas and recreational water environments to the effects of climate change is to apply the DPSIR (Drivers, Pressures, State, Impacts, Responses) framework. This is done in Table 9 (Classification of the impact of climate change on the vulnerability of coastal waters according to the DPSIR approach).

Table 9. Classification of the impact of climate change on the vulnerability of coastal waters according to the DPSIR approach

Drivers	Natural changes in the climate plus all human actions that (might have) affected the changes.
Pressures	Warming of air and water, droughts, rainstorms, flash floods (with special regard to SLR and storm-surges), excessive water abstraction from ground and surface waters. Increasing diffuse-source pollution.
State	State of the quality and quantity of available ground and surface freshwater resources in coastal regions.
Impacts	Saltwater intrusion into fresh groundwater resources: declining, deteriorating fisheries, agriculture, tourism, marine and freshwater ecosystems, health infrastructure, municipal water supply and sanitation systems, deteriorating freshwater quality.
Responses	Revision of water quality standards, improvement of monitoring systems, development of hydrological and hydrogeological models. Actual strategies that were not mentioned in this chapter but may include: • finding alternative freshwater resources, storage of river and runoff water, rainwater harvesting; • development of water desalinating technologies; • building engineering structures that prevent saltwater intrusion into aquifers; • securing much more efficient uses of available freshwater by various non-technical, educational, public information, legal-administrative means, etc.

4.2 Vulnerability of Inland Bathing Waters

Both extremes of a changing climate (extreme drought and rainstorm runoff) may have serious impacts on the quality of inland bathing waters in rivers and lakes, causing a health risk to people bathing there.

Extreme droughts result in lower river discharges and smaller volumes of standing waters and this will, due to reduced dilution, increase concentrations of all types of contaminants discharged into the water. In addition to this, warmer water temperatures will change all temperature-dependent chemical reactions and biological processes, some of which may result in serious deterioration of bathing-water quality. A typical example (illustrated briefly below from Lake Balaton, Hungary) is when increased plant nutrient concentrations, warmer water temperatures and lower water levels (allowing better light penetration) can cause the sudden proliferation of algae, among them the bloom of blue greens, which can have an adverse impact on bathers' health via the toxins they contain. Bathing in freshwater containing toxic cyanobacterial blooms or scum can cause nausea, vomiting, and hay fever-like symptoms, especially in children. The sight and smell of masses of decaying algae would also cause people to avoid the bathing waters.

Drought- and water-scarcity-induced migration of people to the north may also result in the re-emergence in bathing waters of viruses and pathogens that have been long extinct in Europe, such as hepatitis A.

Extreme stormy rainfall runoff may cause sudden and excessive loads of faecal bacteria in bathing waters, especially near urban areas, due to the stormwater overflow of combined sewer systems. Even urban runoff waters arriving via separated sewer systems will cause high loads of many contaminants, including bacteria. Flash storm runoff from agricultural land also sends extra loads of various contaminants into recipient waters and thus to bathing waters via erosion and leaching from rural land, fertilizer- and pesticide-laden agricultural soils, and from various forms of animal husbandry. In Europe the diffuse loads of many contaminants from storm runoff represent the greater part of the total annual load to surface waters (and that was the case even before the onset of the Water Framework Directive (WFD) led to rapidly decreasing point-source sewage and wastewater loads).

Global warming may encourage the appearance of new pathogens. Examples include *Vibrio* species, *Naegleria* and *Acanthamoeba* species that can better proliferate in the warmer waters. New subtropical/tropical cyanobacterial species can invade European bathing waters. Zoonotic infections may also be changed and amplified due to the changes in migrating animal species, especially those of water fowl.

Adaptation and mitigation measures should first be aimed at achieving joint action and data sharing by all levels of water-related authorities, since in most of Europe surface waters belong to both health and water/environmental authorities, which act as separate entities. It is also important to upgrade monitoring systems to include event-based sampling of urban and rural stormwater runoff and the water of the recipient streams and lakes. Data from such monitoring campaigns are practically nonexistent and therefore the actual health risk is not known and reliable counter measures cannot be designed or planned.

Actual adaptation strategies should aim first at developing/applying urban and rural diffuse load reduction technologies.

Table 10. Classification of the impact of climate change on the vulnerability of inland bathing water according to the DPSIR approach

Drivers	Natural changes of the climate plus all human actions that (might have) affected the changes of the climate.
Pressures	Warming of air and water, drought, rainstorms, flash floods. Increasing point (end-of-pipe) and diffuse-source pollution.
States	State of the quality and quantity of bathing waters of freshwater lakes and streams, including their ecological status.
Impacts	Increasing concentrations of all pollutants due to less dilution by water in drought-affected regions. Excessive growth of algae, including blue greens. Warming may cause the appearance of new pathogens. In the case of flash rainstorm runoff, sudden and excessive loads of faecal bacteria in bathing waters can be expected, especially near urban areas. High loads of many pollutants from agricultural land caused in this way also adversely affect bathing-water quality. Changes in the migration of water fowl may be the source of extra bacterial loads to bathing waters (especially in lakes).
Responses	Health and water/environmental agencies should merge their databases and harmonize monitoring programmes. The latter must focus on storm/runoff events in both runoff water and in bathing water to provide data for planning adaptation and preventative strategies. Concrete adaptation and control strategies should aim not only at developing and applying urban and rural diffuse load reduction technologies (the BAT, Best Available Technology) but also at developing novel solutions and at the careful regulation of lake water levels (when this is an option).

Among these, the provision of storage of both urban and rural storm runoff water by appropriately designed polishing ponds and wetland systems is the most urgent task. On agricultural land this should be accompanied by the familiar techniques of keeping and storing water in the soil (contour line tillage, terracing, vegetation buffer strips, etc.) thus also removing contaminants and preventing erosion.

More information on the contamination of bathing waters can be found at the web site of the Hungarian strand information (http://www.strandinformacio.hu/index.php?lang=en). The value of applying the DPSIR approach is shown in Table 10 (Classification of the impact of climate change on the vulnerability of inland bathing water according to the DPSIR approach).

4.3 Saline Water Intrusion in Aquifers Used for the Production of Drinking Water

Global climate change and extreme weather events (e.g. severe storms, droughts and floods) are expected to have negative effects on the quantity and quality of water resources (EEA, 2007), amplifying the anthropogenic pressures on surface water and groundwater resources (Hiscock & Tanaka, 2006) resulting from the growth in the global population and the demand for potable water. Groundwater is a key source of supply, especially of drinking-water. In fact, worldwide more than 2 billion people depend on groundwater for their daily consumption (Kemper, 2004).

From a water quality perspective, extreme dry periods may result in a lower dilution capacity and in increased concentration of contaminants in groundwater aquifers, especially in the unconfined ones. In quantitative terms, reduced groundwater recharge during dry periods and increased water abstraction due to warmer temperatures may cause further water stress by reducing groundwater table levels.

Particularly in coastal areas where the pressure on water demand is very high due to population density, agriculture and tourism, the consequences of extreme weather events on freshwater aquifers may be exacerbated by the intrusion of seawater into freshwater aquifers.

Larger storm surges produced by extreme storms, combined with a rising sea level, could result in much higher rates of coastal erosion, which would in turn affect the levels of saline intrusion into coastal freshwater (OzCoast, 2010).

Coastal freshwater aquifers are strategic resources that provide water for many important uses including town water supply, domestic water supply, irrigation of crops and industrial processes. In addition to the threats posed by extreme events and rising sea levels, saltwater intrusion into groundwater could be exacerbated by human abstraction and over-exploitation.

In general freshwater contamination by seawater of only 5% is usually enough to preclude many important uses including drinking-water supply, irrigation of crops, parks and gardens, and the well-being of groundwater-dependent ecosystems (UNSW, 2010). So the prevention of saltwater intrusion into coastal aquifers is a key challenge, not only to maintain adequate water quality for human consumption, but also to permit other possible human uses of groundwater, particularly under predicted climate change and extremes.

At international and European level, water quality standards are used to protect human health from the adverse effects of saltwater contamination of drinking-water. Limit values of sulphate/chloride/conductivity are useful for estimating seawater intrusion and the intrusion of non-seawater salts into groundwater (Council of the European Union, 2006 (Directive 2006/118/EC)). The Drinking-water Directive (DWD) (Council of the European Union, 1998 (Directive 98/83/EC)) fixed a salinity limit, measured as conductivity, equal to 2500 µS/cm.

According to WHO (2006) no health-based guideline value for chloride in drinking-water was proposed (an excess of 250 mg/litre can give rise to a detectable taste in water). WHO stated in 2006 that the average daily intake of sulphate from drinking-water, air and food is approximately 500 mg, food being the major source. The existing data do not identify a level of sulphate in drinking-water that is likely to cause adverse human health effects. However, because of the gastrointestinal effects resulting from ingestion of drinking-water containing high sulphate levels, it is recommended that health authorities be notified of sources of drinking-water that contain sulphate concentrations in excess of 500 mg/litre.

Important quality standards are also available to set the suitability of groundwater quality for irrigation and protection of plants from saltwater contamination. According to Camberato (2001) irrigation water is classified under four categories based on the salinity hazard, which considers the potential for damaging plants and the accompanying management measures needed for use as an irrigation source (see Table 11 (Classification of saline irrigation water).

In addition to water quality standards, groundwater monitoring systems are necessary, particularly in coastal areas where it is important to check saltwater contamination. The implementation of monitoring programmes to protect drinking-water quality is an important task. The use of adequately designed monitoring networks can help in optimizing the number of sampling points, including choosing adequate times and suitable sampling positions and constructing an efficient and optimized sampling network (Marangani, 2008). Another tool to help managers in detecting seawater intrusion and monitoring future contamination of coastal aquifers is the application of integrated indexes.

According to Edet & Okereke (2001), monitoring should be based on an Assessment Index (AI) considering the following indicators of saltwater intrusion: total dissolved solids (TDS), density (D), sodium (Na), chloride (Cl), and Br/Cl ratio. Additionally, in order to characterize the complexity of salinization and groundwater evolution processes, and to determine the spatial extension of the saltwater contamination, DiSipio, Galgaro & Zuppi (2006) proposed the use of geochemical and isotopic analyses associated with electrical conductivity data.

Generally, hydrogeological exploration programmes for the assessment of salt intrusion processes require the use of dedicated monitoring wells and piezometers, the collection of water samples and aquifer performance/water quality tests. The use of multiple-depth monitoring wells including several piezometers installed at different depths can help in monitoring groundwater levels and water quality and determine whether the hydraulic gradient is established from the coast to the pumping wells, indicating possible seawater intrusion (Danskin & Crawford, 2008).

In order to evaluate the potential impacts and risks of seawater intrusion into coastal aquifer systems and to employ appropriate monitoring systems and adaptive measures, especially during extreme weather events, it is essential to characterize both exposure to climatic factors and the sensitivity of groundwater resources to climate variations and extremes (see also Chapter 2).

A comprehensive knowledge of the nature of climatic variation in space and time is vital to characterize exposure and therefore climatic stressors that impact on a system. The combination of hydrological and hydrogeological models with climate

Table 11. Classification of saline irrigation water

Salinity class	Electrical conductivity µS/cm	Total dissolved salts (ppm)	Potential injury and accompanying management measures
Low	<250	<150	Low salinity hazard; generally not a problem; additional management is not needed.
Medium	250–750	150–500	Damage to salt-sensitive plants may occur. Occasional flushing with low-salinity water may be necessary.
High	750–2500	500–1500	Damage to plants with low tolerance to salinity will likely occur. Plant growth and quality will be improved with excess irrigation for leaching, and/or periodic use of low-salinity water and good drainage provided.
Very high	>2500	>1500	Damage to plants with high tolerance to salinity may occur. Successful use as an irrigation source requires salt-tolerant plants, good soil drainage, excess irrigation for leaching, and/or periodic utilization of low-salinity water.

Source: USDA (1954).

models gives the opportunity to adjust global climate model (GCM) bias and include improved representations of hydrological processes. Moreover, a combination of statistical techniques and regional modelling can be used to downscale climate models and to simulate weather extremes and variability in future climates.

So regional hydrological and groundwater models integrated with climate and statistical models allow investigators to consider the groundwater balance in the aquifer and to simulate saltwater intrusion under present and future scenarios of climate change. Finally, together with integrated modelling, empirical approaches such as the "analogue approach" give information that is more specific than that given by the GCMs. By reconstructing past climates (i.e. temperature and precipitation) in a given area these approaches can be used to construct future scenarios by analogy (Dragoni & Sukhija, 2008).

4.4 Consequences of Extreme Weather Events for Bathing-Water Quality

Vulnerability has to be taken into account from two perspectives – the susceptibility of i) people and ii) the water source. Personal vulnerability is critical for elderly or immunocompromised people. Also, with respect to diarrhoeal diseases, young children can be considered as a highly susceptible group.

The vulnerability of water sources plays a massive role in changing bathing-water quality. Shallow lakes, for example, are more vulnerable to droughts than deep ones (Case study 1). At lower water levels, or with decreased water volumes, concentrations of all pollutants increase, including plant nutrients, and the water temperature rises more rapidly, resulting in better conditions for emergent tropical and subtropical pathogens to proliferate

In several European countries, especially in the Nordic countries, enteric viruses, mainly Norovirus infections, remain a major waterborne risk (Risebro et al., 2007). It is likely that the risk of disease from these infections elsewhere in Europe is much greater than currently realized. The large majority of enteric viral infections is strictly anthropogenic, but climate change impacts could bolster the (re-)emergence of viral infections as a result of poorer quality water because of seasonal drying-out or extreme weather.

In addition to these demographic changes, migration into Europe as a consequence of climate change elsewhere could bring in new viruses and reintroduce those that have declined in importance in the continent. For example, viral hepatitis A infection (HAV) has decreased substantially in most European populations in recent decades as a result of improved sanitation and hygiene.

But immigration into Europe from Africa will lead to an increased incidence of HAV in immigrant populations, and this could lead to an increased risk of re-emergence, including its waterborne spread. Climate change will also affect avian and mammal populations which could bring in new viruses of risk to health. For example, although primarily a respiratory disease, SARS (severe acute respiratory syndrome), a zoonotic infection, was shown in one outbreak to spread through wastewater.

4.5 Water Quality Changes Caused by Extreme Weather

4.5.1 Stormy rainfalls

Heavy rainstorms causing increased flood runoff, erosion and the washing-off of large amounts of contaminants will have a major impact on the abundance of waterborne pathogens in bathing waters because of increased agricultural soil erosion, the overflow of rural and urban sewage treatment systems, and chiefly because of urban stormwater runoff load. A lot of faecal pathogens will also be washed into bathing waters.

Zoonotic infections may expand as the faeces of wild birds and mammals (notably rodents) on the beaches are washed into

Case Study 1: Consequences of Drought in a Shallow Lake (Lake Balaton), Hungary, 2003

Lake Balaton is situated in the western part of Hungary, in Transdanubia. It is the largest shallow lake in central Europe with a surface area of 593 km^2. The average depth of the lake is 3.14 m and 11.0 m at the deepest point. As a consequence of a series of dry years at the turn of the millennium, the water level of the lake dropped in 2003 to 23 cm below the lowest regulation level, which is 70 cm lower than the multi-annual minimum level. The average annual precipitation into the drainage basin of the lake was 507 mm in these dry years, which is 110 mm lower than the multi-annual mean.

Altogether, the depth of the water off the southern flat beaches of the lake turned to as low as 10–20 cm along hundreds of metres from the shoreline towards the middle of the lake. As shown by the photographs on p. 38, green algal blooms occurred along all the beaches. This algal material concentrated the diatoms into a mass that caused skin irritation by their lance-like structure. A lot of complaints about this were registered during this period. In addition, the sight and smell of algal scum were repulsive, frightening visitors away from the beaches.

Source: compiled by representatives of the National Institute of Environmental Health, Hungary.

the water. Increased erosion from agricultural areas may increase nutrient loads in bathing water, providing better conditions for the proliferation of toxic cyanobacteria. Bathing in freshwater containing toxic cyanobacterial blooms or scum can cause nausea, vomiting and hay fever-like symptoms, especially in children, because they play in shallow water where the blooms can accumulate. The amount of cyanotoxins can reach the daily tolerable intake (DTI), especially in children.

4.5.2 Global warming

Warmer air temperature may warm the water and new pathogens or harmful species can appear which are new to the European region. Examples include *Vibrio*, *Naegleria* and *Acanthamoeba* species. New subtropical/tropical cyanobacterial species can also invade European bathing waters. But the survival rate of some bacteria may be decreased by higher ultraviolet (UV) irradiation.

4.5.3 Droughts and water scarcity

Droughts or water shortages can affect bathing-water quality because the decreased stream flows are not sufficient to dilute sewage and wastewater loads. This results in an increase in the concentration of pathogens and can cause infections in a greater number. As Case study 1 shows, the decreased volumes of standing water also result in increased concentrations of contaminants such as plant nutrients, and may lead to several water quality problems, including the increased frequency of cyanobacterial blooms.

Algal contamination in Lake Balaton

Mechanical removal of algal beds

© National Institute of Environmental Health, Hungary

Close contact of bathers with algae can pose a direct health risk

4.6 Elements of Mitigation Measures for Bathing Waters

4.6.1 Joint information systems and exchange of information

In most of the Member States in the WHO European region the database of chemical and ecological water quality is the responsibility of environmental ministries or agencies. But the microbiological data of bathing-water quality, including the pathogens' occurrence, belong to health ministries or networks. In extreme weather situations it is absolutely essential for all agencies to share data and to carry out field measurements together.

4.6.2 Prevention of stormwater overflow at sewage treatment plants

In planning and constructing sewage water treatment plants the overflow of untreated sewage must be prevented during heavy rainstorms.

4.6.3 Prevention of erosion and diffuse pollution by appropriate land-use techniques

Erosion of nutrient-rich and fertilized agricultural soils must be prevented, along with the reduction of runoff water that contains high concentrations of nutrients and other contaminants. Similarly diffuse- or non-point source control measures should be carried out in urban and populated rural areas. The techniques available for keeping polluted washoff away from bathing waters vary very much, and several books have been published on this subject (e.g. Thornton et al., 1999). The best-known techniques for agricultural land include contour line tillage, strip cropping, vegetation buffer strips and terracing.

4.6.4 Monitoring during extreme weather events and risk assessment

Monitoring systems must be upgraded to include the sampling of extreme weather-induced events, both in the runoff water and in the recipient water bodies, by all health and water/environmental agencies involved. Without such event-based measurements a risk assessment of bathing waters cannot be made.

4.6.5 Public awareness and information

One of the main issues to handle in relation to extreme weather events is to inform the public in good time about the dangers and risks of the situation that is expected. It can be done via the Internet and/or by other media.

Chapter 5

IMPACTS OF CLIMATE CHANGE AND EXTREME EVENTS ON WATERBORNE DISEASES AND HUMAN HEALTH
Bettina Menne
WHO Regional Office for Europe

Health workers in front of a hospital in Muynak, Uzbekistan in March 2008
© WHO/Tanja Wolf

Impacts of Climate Change and Extreme Events on Waterborne Diseases and Human Health

5.1 Key Messages

Temperature increases, rainfall fluctuations, periods of droughts and heat waves, as well as SLR have a significant potential to affect freshwater resources, wastewater and land-related processes, and through this affect human health.

- Drought and severe weather can favour infectious disease outbreaks and impair water quality, hygiene and sanitation.

- Heat waves may imply restrictions and prioritization of water use, control of drinking-water quality and compromised efficiency of sanitation systems.

- Cold spells may affect water, electricity and heating system availability, with potential impacts on population health and health service delivery.

- Flooding has direct health effects such as drowning, injuries, diarrhoeal diseases, vector-borne diseases, respiratory infections, skin- and eye infections, and mental health problems.

- Heavy rainfall/floods can cause over-flooding of sewage treatment plants, runoff of animal dejections and manure with consequent increase of contamination of surface water and soil.

- Land use in water catchment areas is gaining in importance to assess and manage risks to human health.

- Rains lead to higher concentrations of pathogens in the aquatic environment, affecting the quality of bathing water, drinking-water resources, and potentially some foods such as aquatic and aquaculture products. Heavy rains and floods can also increase the nutrient availability of lakes, inducing cyanobacterial proliferation.

- Ecosystem changes may allow colonization by pathogens of hitherto unaffected environments.

Extreme weather events will pose new challenges to the protection and promotion of human health. Current safeguards will need to be assessed, and technologies capable of adapting to the range of climate change scenarios need to be identified and prioritized.

Table 12. Observed and projected changes in climate conditions: potential risks and opportunities

Climate change risks	Potential risks and opportunities		
	Freshwater resources	Wastewater	Land-related processes
Increase in summer temperatures	Increased demand for potable water, increased pressures on groundwater, increased demand on reverse osmosis plants, increased evapotranspiration rates	Increased sewer dry weather flow, increased dry weather treatment volumes, increased treated effluent volumes	Reduction in groundwater recharge, more aggressive regime for agriculture Ground shrinkage
Increasing winter temperatures	Increased demand for potable water, increased pressures on groundwater, increased demand on reverse osmosis plants, increased evapotranspiration rates	Increased sewer dry weather flow, increased dry weather treatment volumes, increased treated effluent volumes	Productive regime for agriculture with opportunities for premium products maturing early
Higher winter rainfall	Increased volumes for recharge, existing water storage volumes might be insufficient, increased stormwater runoff	Higher volume of stormwater generated which may exceed infrastructure capacity, higher volumes of stormwater entering sewers – surcharge events increase, increased volumes of wastewater to treat at sewage treatment plants, increased volumes of treated effluent may remain unutilized	Increased flooding incidence, increased damage to infrastructure, increase in soil erosion
Lower summer rainfall	Lower recharge volumes increase in demand from agricultural sector	Lower sewage volumes and consequent treated effluent volumes	Ground shrinkage
Higher intensity of rainfall	Higher proportion of total rainfall might end up as runoff and not contribute to recharge volumes Higher level of pollutants in stormwater	Higher peak flows in sewers, increased possibility of sewer surcharge and overflows	Increased incidence of flooding damage to infrastructure, increased soil erosion
SLR	Reduced volumes of groundwater, increased salinity of groundwater	Increased seawater infiltration volumes, more saline wastewater and hence treated effluent	Loss of land, increased flooding of coastal areas, increased need for flood defences, new methods of construction, insurance premiums may increase

Source: Gatt (2009).

5.2 Lower Rainfall and Drought

The IPCC projections (IPCC, 2007) illustrate decreasing water availability and increasing frequency of drought in mid-latitudes and semi-arid low latitudes, with increases in drought frequency for southern and south-eastern Europe, as well as central and south-east Asia. Observed and projected changes in climate changes in climate conditions present potential risk and opportunities. These are summarized in Table 12. Droughts and severe water stress have major effects on various sectors such as agriculture, forestry and industry. Droughts can damage ecosystems and increase the risk of wildfires. In southern Europe, climate change is projected to exacerbate conditions (through high temperatures and drought) in a region already vulnerable to climate variability, and to reduce water availability, hydropower potential, summer tourism and, in general, crop productivity (IPCC, 2007).

The impacts on water management and human health are various: decreased river flows with increased concentrations of pathogens; less water for dilution of sewage effluent discharges; intrusion of organic material along the distribution network when system pressure drops significantly; intermittent piped water supply with a risk of introduction of contaminants; reduced water supply; lowering of groundwater table in coastal areas (lower recharge and excessive withdrawals) and increasing seawater intrusion which can lead to salinization of available water resources; and the increased use of wastewater in agriculture (Menne et al., 2008; Frangano et al., 2001; Senhorst & Zwolsman, 2005). Each of these conditions can favour infectious disease outbreaks and impair water quality, hygiene and sanitation. Droughts can also affect the transmission of some mosquito-borne diseases (Bouma & Dye, 1997; Woodruff et al., 2002; Chase & Knight, 2003; Githeko et al.,

2000). Drought may exert an influence on cyanobacterial proliferation by increasing nutrient availability (higher concentrations due to surface water evaporation in summer) and reducing water body flow (thus increasing the area of still waters which promote their growth).

The effects of drought on health include, besides these, deaths, malnutrition (undernutrition, protein-energy malnutrition and/or micronutrient deficiencies), infectious and respiratory diseases (Menne & Bertollini, 2000). Drought and the consequent loss of livelihoods is also a major trigger for population movements, particularly rural-to-urban migration. Population displacement can lead to increases in communicable diseases and poor nutritional status resulting from overcrowding, and a lack of safe water and food (Menne & Bertollini, 2000; Del Ninno & Lundberg, 2005).

5.3 Heat Waves

The knowledge of health impacts from extreme temperature and linkages to water supply during periods of very hot and dry weather and possible health effects may imply restrictions and prioritization of water use, control of drinking-water quality, efficiency of sanitation systems and a requirement for collaboration between the health sector and water suppliers.

Episodes of extreme temperature have affected health significantly. For instance, in the summer of 2003, a severe heat wave struck much of western Europe. Twelve European countries reported more than 70 000 deaths above the average of the five previous years (Robine et al., 2008). For populations in the EU, mortality has been estimated to increase 1–4% for each degree increase of temperature above a cut-off point (Menne et al., 2008). The PESETA (Projection of Economic impacts of climate change in Sectors of the European Union based on bottom-up Analysis) project (PESETA, 2008) estimates 86 000 extra deaths per year in EU countries with a global mean temperature increase of 3 °C in 2071–2100 relative to 1961–1990. Increasing numbers of older adults in the population will increase the proportion of the population at risk. Heat waves have larger effects on mortality when air pollution is high.

During heat waves, water as well as electricity consumption/demand increases, sometimes together with a water stress-related decrease of hydropower potential. Hot and dry weather often coincides with lack of water or rainfall, and thus over an extended period of time may also influence water quality as outlined above.

5.4 Higher Water Temperatures

It is interesting that pathogens in the water environment are typically less easily neutralized by lower temperatures: higher temperatures would exert selection towards less temperature-sensitive species, directly promoting the growth of some indigenous bacteria, including pathogenic species (Lipp, Huq & Colwell, 2002; Kirshner et al., 2008). In particular, enteric bacterial pathogens are not able to replicate in the aquatic environment. They are inactivated in the aquatic environment at a rate that increases with temperature. The sensivity to temperature of enteric pathogens varies: cysts of *Giardia* and enteroviruses are less rapidly inactivated compared with oocysts of *Cryptosporidium* (Schijven & De Rosa Husman, 2005). It is also known that a large variation in temperature susceptibility exists among viruses (Schijven & Hassanizadeh, 2000). For instance, viruses like the HAV are fairly temperature-insensitive. Ultraviolet light, too, decreases survival. Increased temperature and increased sunlight will increase the rate of die-off of pathogens in the environment.

5.5 Cold Spells

Cold weather still threatens the health of many European populations. Most European countries have between 5% and 30% higher death rates in winter than in summer. People with cardiovascular diseases are more at risk in winter, because of the cold-induced tendency for blood to clot. However, overall winter mortality rates are falling in some European countries. Cold spells may affect water, electricity and heating system availability, with potential impacts on population health and health service delivery (Case study 2). They may also affect transport and therefore slow down access to health services.

5.6 Higher Rainfall, More Intense Rainfall and Floods

Flooding is the most common natural extreme weather event in the European region (EM-DAT, 2009). With climate change, winter floods are likely to increase throughout the region. Coastal flooding related to increasing frequencies and intensities of storms and SLR is likely to threaten up to 1.6 million additional people annually in the EU alone.

Case Study 2: Cold Spell in Tajikistan, 2008

The recent cold spells in central Asia gives an example of possible health consequences. In 2008, Tajikistan had the coldest winter for 30 years, with electricity generation impaired through frozen inlet streams. Consequently health services and households had no energy for prolonged periods.

A rapid health assessment showed a sharp increase in the number of cases of severe burns and frostbite, a 50% increase in hospital admissions from acute respiratory infections, and a doubling of maternal and infant mortality compared with the same period in 2007.

Source: compiled by representatives of WHO Regional Office for Europe.

The potential health effects of flooding include (Vasconcelos, 2006) direct health effects: drowning, injuries (cuts, sprains, lacerations, punctures, electric injuries, etc.), diarrhoeal diseases, vector-borne diseases (including those spread by rodents), respiratory infections, skin and eye infections, and mental health problems; and other effects with health consequences: damage to infrastructure for health care and water and sanitation, crops (and/or disruption of food supply) and property (lack of shelter), disruption of livelihoods, and population displacement.

The limited data available on flood events from a few event-based epidemiological studies show that the greatest burden of mortality is from drowning, heart attacks, hypothermia, trauma and vehicle-related accidents in the immediate term (Meusel et al., 2004). Studies of the long-term health effects of floods are lacking (WHO Regional Office for Europe, 2005).

Flooding may lead to contamination of water with dangerous chemicals, heavy metals, or other hazardous substances, from storage or from chemicals already in the environment (e.g. pesticides) (Pardue et al., 2005). Unfortunately there is little published evidence demonstrating a causal effect of chemical contamination on the pattern of morbidity and mortality following flooding events (Euripidou & Murray, 2004; Ahern et al., 2005). Heavy rainfall/floods can cause over-flooding of sewage treatment plants, runoff of animal dejections and manure with a consequent increase of contamination of surface water and soil. Land use in water catchment and drainage areas will become progressively more important for inclusion in risk assessment and risk management.

Several studies have shown that contamination of freshwater by enteric pathogens is higher during the rainy season (Nchito et al., 1998; Kang et al., 2001) (Case study 3). In a study in the Netherlands (Schijven & De Rosa Husman, 2005), one of the main conclusions was that increased precipitation in winter and more frequent heavy rain in summer lead to peak concentrations of waterborne pathogens in surface waters that are several orders of magnitude above average levels. These higher pathogen concentrations can affect the quality of drinking and bathing waters, and some foods like vegetables, soft fruits and shellfish. Moreover, the increased load of suspended particles can jeopardize the efficiency of water filtration and treatment systems, increasing the risk of drinking-water supply contamination. Beyond that, many waterborne disease outbreaks are related to heavy precipitation events, often linked to water treatment failures (Bates et al., 2008; Kistemann et al., 2002). For example *Cryptosporidiosis* cases in England and Wales between April and July were positively associated with maximum river flow (Lake et al., 2005). In Europe, outbreaks are frequently preceded by heavy rainfall (Miettinen et al., 2001). However, it is not feasible to extrapolate the impacts of these events in terms of climate change (McMichael et al., 2004). It is worth noting that heavy rain and floods can increase the nutrient availability of lakes, so inducing cyanobacterial proliferation. Furthermore, increased freshwater inputs into rivers from increased runoff can dilute the estuarine environment and promote blooms of toxic cyanobacteria. For example, a large bloom of *M. aeruginosa* in the upper San Francisco Bay Estuary was widespread throughout 180 km of waterways; MCs were detected at all stations sampled, and also in zooplankton and clam tissues (Fristachi & Hall, 2008).

Overall, the risk of infectious disease following flooding is generally low, although increases in diarrhoeal diseases after floods have been reported (Miettinen et al., 2001; Reacher et al., 2004; Wade et al., 2004; Wolf & Menne, 2007).

5.7 Changes in Ecosystems

Changes in the phenology (periodic biological phenomena) and the distribution of marine species have been observed, such as earlier seasonal cycles and northward movements which will change marine ecosystems and biodiversity and affect fisheries. Changes in ecosystems will also affect the spread of water-related diseases. Increasing temperatures and changing water quality may create new ecological niches that allow pathogens to invade new areas (Case study 4). Recreational water quality may be degraded by changing natural ecosystems or decline in the quality of waters draining to bathing areas. This may increase the risk of infection for bathers, as well as affecting seafood.

5.8 Changes in Seasonality

Water- and foodborne outbreaks commonly follow seasonal patterns, and hence are likely to be affected by changes in climate. For example, illness related to campylobacter and salmonella exhibit a summertime pattern of occurrence (Greer, Ng & Fisman, 2008). *Cryptosporidium* concentrations are highest in waterways in spring during the calving and lambing season.

5.9 Changes in Human Behaviour

Climate change will also affect people's habits. Warmer temperatures are expected to increase the consumption of fruit, salads, vegetables and drinking-water. Warmer, drier conditions will increase recreational water use, potentially increasing the length of the bathing season.

5.10 SLR

According to satellite observations, the pace of global mean SLR has increased to more than 3 mm/year in the last decade (as compared to a global average in the 20th century of 1.7 mm/year). Because of ocean circulation and gravity effects SLR is not uniform but varies across European seas. Sea-surface temperature increases have also accelerated in recent decades. For the future, projections suggest European sea level and sea-surface temperature will rise more than the global average, with significant impacts on human well-being and ecosystems in coastal regions. IPCC SLR estimates (IPCC, 2007) are possibly conservative because of the risks of more rapid changes in the Greenland (and Antarctic) ice sheets than assessed so far.

Case Study 3: Environmental Health Aspects of Flooded Karstic Drinking-water Resources, Hungary

This case study describes a large-scale drinking-water disease outbreak. The water supply of a Hungarian city is mainly based on the sensitive karst water springs. An enormous quantity of rain fell on the catchment area of the water source, causing an unusually strong water flow and flooding. Microbiological contamination from several potential sources in the protecting zone of the water was washed into the wells and water mains and caused an outbreak: 3673 people became ill out of the 60 000 living in the water supply zone, and 161 were admitted to hospital. Public health intervention and hygienic measures were made in line with epidemiological actions, and focused on:
i) the protection of the healthy people through providing safe drinking-water
ii) the identification of the contamination; and
iii) taking measures on risk reduction and preventive actions.

Source: compiled by representatives of the National Institute of Environmental Health, Hungary.

Case study 4: Changes in the Marine Food Web in Europe

Northeast Atlantic Ocean

Projections indicate that warming will extend throughout the water column during the course of the 21st century (Meehl et al., 2007). Sea-surface temperature changes have already resulted in an increased duration of the marine growing season and a northward movement of marine zooplankton. Some fish species are shifting their distributions northward in response to increased temperatures.

Baltic Sea

Climate models project a mean increase of 2–4 °C in the sea-surface temperature in the 21st century, and increasing run-off and decreasing frequency of Atlantic inflows, both of which will decrease the salinity of the sea. Consequently, the extent of sea ice is expected to decrease by 50–80% over the same period (Meier, Kjellström & Graham, 2006) and stratification is expected to become stronger, increasing the probability of a deficiency of oxygen (hypoxia) that kills a lot of marine life in the region. Changes in stratification are expected to affect commercially important regional cod fisheries because stratification appears to be an important parameter for the reproductive success of cod in the Baltic Sea.

Mediterranean Sea

Temperature is projected to increase and runoff to decrease. In contrast to the Baltic Sea, the combination of these two effects is not expected to change stratification conditions greatly because of the compensating effects of increasing temperature and increasing salinity on the density of sea water. The invasion and survival of alien species in the Mediterranean is correlated with the general sea-surface temperature increase, resulting in the replacement of local fauna with new species. Such changes affect not only local ecosystems, but also the activities of the international fishing fleet when commercial species are affected (Phillippart, 2007).

Source: compiled by representatives of ISPRA, based on EEA (2008).

SLR can affect human health through coastal degradation, higher tidal surges and coastal flooding.

5.11 Climate Change and Diarrhoeal Diseases

Estimates of the burden of diarrhoeal disease attributable to water, sanitation and hygiene in the European region for children 0–14 years of age amount to over 13 000 deaths (5.3% of all deaths in the 0–14 age group), mostly coming from countries of eastern and south-eastern Europe and central Asia (Valent et al., 2004). This is equivalent to 23% of the population being affected by episodes of diarrhoea each year, although the rate varied across different zones, from 19% in Eur-A countries to 36% in Eur-B and 20% in Eur-C. These cases of diarrhoeal disease have a high associated cost in terms of lost working time and health care (Laursen et al., 1994). Campbell-Lendrum and colleagues (2003) estimated the number of additional cases of diarrhoea due to temperature increases through to 2030. A dose–response relationship of 5% increase per degree was used for developing countries, but a conservative estimate of 0% increase per degree was assumed for developed regions. The authors however consider that taking into account the relationship between pathogen behaviour and temperature (e.g. Kovats, Hajat & Wilkinson, 2004) it is likely that additional cases of diarrhoeal diseases will be experienced.

Table 13 (Projected incident cases under high and low emission scenarios by 2030) gives an idea of the possible increase in cases by 2030.

5.12 Some Specific Examples of Climate Change and Waterborne Diseases

Pathogenic *Vibrio* spp., such as *Vibrio parahaemolyticus* and *V. vulnificus*, occur in estuarine waters throughout the world and are present in a variety of seafood (Croci et al., 2001; De Sousa et al., 2004; DePaola et al., 1990, 2003). They are part of the natural flora of zooplankton and coastal fish and shellfish. Their number is dependent on the salinity and temperature of the water, and cannot be detected in water

Table 13. Projected incident cases under high and lower emission scenarios by 2030

Subregion	Climate	2000		2030	
		Mid	High	Mid	High
Eur-A	S550 low	0	1 584	0	4 753
	S750 high	0	1 584	0	4 753
	UE	0	1 584	0	6 338
Eur-B	S550 low	785	2 355	785	5 496
	S750 high	785	2 355	785	6 281
	UE	785	2 355	785	7 066
Eur-C	S550 low	958	1 437	0	3 352
	S750 high	958	1 437	0	3 352
	UE	958	1 915	0	3 831

Source: Ebi (2008).
Notes: For details of country groupings Eur-A, Eur-B and Eur-C, visit the web site of WHO (http://www.who.int/choice/demography/euro_region/en/print.html, accessed 20 September 2010). Medium and high levels differ in the confidence level taken into account.
The table shows three scenarios: low emission scenario stabilization at 550 ppm CO_2, high emission scenario stabilization at 750 ppm CO_2 equivalent, and unmitigated emissions (EU).

with a temperature below 15 °C. With the possibility of acquisition of virulence genes by environmental strains and with a changing climate, the geographic range of these pathogens may also change, potentially resulting in increased exposure and risk of infection for humans. Indeed *V. parahaemolyticus* and *V. vulnificus* are responsible for the majority of the non-viral infections related to shellfish consumption in the United States, Japan and south-east Asia (Wittman & Flick, 1995), and they also occur occasionally in other parts of the world. To date the number of cases which have occurred in Europe is extremely low, but recently a large outbreak (64 cases) was registered in Spain, caused by consumption of *V. parahaemolyticus*-contaminated shellfish harvested in Galicia (Lozano-Leon et al., 2003). The FAO/WHO risk assessment of *V. vulnificus* in raw oysters found a *V. vulnificus*–temperature relationship in oysters at harvest (FAO & WHO, 2005). In the Mediterranean Sea the emergence of *V. vulnificus* has been attributed to higher water temperatures, leading to an increased risk of systemic vibriosis from handling or consuming seafood (Paz et al., 2007). (See Table 14 (Summary table on pathogens and health significance) for further details.)

Furthermore, changes in plankton populations and other hosts for which *Vibrio* spp. are commensals or symbionts would similarly alter their ecology. Indeed, an increased production of exudates from algal and cyanobacterial proliferations are expected to further promote the growth of autochthonous pathogens (Lipp, Huq & Colwell, 2002; Eiler et al., 2007), and an increased presence of *Vibrio* spp. (including the serotypes of *V. cholerae* O1 and O139, and *V. vulnificus*) has been frequently associated with blooms of cyanobacteria and eukaryotic phytoplankton species (Epstein 1993; Eiler et al., 2007).

V. cholera is considered a model for understanding the potential for climate-induced changes in the transmission of foodborne disease, since peaks of disease are seasonal and associated with higher water temperature and phytoplankton (FAO, 2008). FAO and WHO have undertaken a risk assessment of choleragenic *V. cholera* O1 and O139 in warm water shrimp for international trade.

Based on the available epidemiological data, it was estimated there was a very small risk of acquiring cholera through consumption of imported warm water shrimp. But further research needs to be conducted to address the existing data gaps (FAO & WHO, 2005). The EU Rapid Alert System for Food and Feed (RASFF) has had several alerts in recent years related to *V. cholera* in imported warm water shrimps from different countries (European Commission DG Health and Consumer Affairs, 2007). In 2007, the product category for which the most RASFF alerts were sent was fish, crustaceans and molluscs (21%). These alerts have a consequent socioeconomic impact associated with the withdrawal and/or recalls of the seafood products (European Commission DG Health and Consumer Affairs, 2007) and consequent risk perception.

Another important group of autochthonous pathogens belongs to the genus *Legionella* which can cause a number of infections of differing severity, generically referred to as legionellosis, ranging from mild febrile illness (Pontiac fever) to a potentially fatal form of pneumonia.[9] The presence of Most

---

[9] *Legionella pneumophila* is the well-known agent of this pneumonia, especially fatal in elderly people. Outbreaks of legionellosis have been reported from all countries in Europe, many of them linked to hotels and other types of holiday accommodation or to those systems where water temperature is higher than ambient temperature. Legionellosis is indeed caused by aerosols from contaminated potable water distribution systems, cooling towers, building water systems, respiratory therapy equipment and hot tubs, but the presence of *Legionella* spp. in these artificial water systems depends ultimately on the reproductive success of the bacterium in the natural environment (Bartram J, 2007).

Table 14. Summary table on pathogens and health significance

	Pathogen	Weather influences	Health significance[b]	Relative infective dose[b]	Infection caused
Viruses	Norovirus GGI and GGII	Storms can increase transport from faecal and wastewater sources	High	Low	Gastroenteritis
	Sapovirus		High	Low	Gastroenteritis
	HAV		High	Low	Hepatitis
	Rotavirus		High	Low	Gastroenteritis
	Enterovirus		High	Low	Gastroenteritis
	Adenovirus		High	Low	Respiratory & intestinal influenza
	Avian influenza virus[a]		Low	Unknown	
		Survival increases at reduced temperatures and sunlight (ultraviolet)			
		Changes in seasonality			
Bacteria	Pathogenic *Escherichia coli*	Enhanced zooplankton blooms	High	Low	Gastroenteritis
	Campylobacter jejuni, C. coli		High	Moderate	Gastroenteritis
	Helicobacter pylori		High	Unknown	Stomach & duodenal ulcer
	Legionella pneumophila				Pneumonia
	Vibrio cholerae	Salinity and temperature associated with growth in marine environment	High	High	Cholera
	Vibrio parahaemolyticus[a]		High	High	Wound infections, otitis and lethal septicaemia, gastroenteritis, respiratory dysfunctions, allergic reactions
	Vibrio vulnificus[a]				
	Vibrio alginolyticus		Medium	High	
	Toxic cyanobacteria		Low	Low	
			Low	Unknown	
			Medium	Moderate	
Protozoa	*Cryptosporidium* spp.	Storms can increase transport from faecal and wastewater sources Temperature associated with maturation and infectivity of Cyclospora	High	Low	Gastroenteritis
	Giardia spp.		High	Low	Gastroenteritis
	Naegleria fowleri[a]		Low	High	Meningoencephalitis
	Acanthamoeba spp.[a]		Low	Unknown	Keratitis, blindness

Source: Pond et al. (unpublished).
Notes: Taxa labelled with [a] are considered potentially emerging; [b] according to WHO (2003).

Legionella spp. are spread over a wide range of natural environmental conditions, with better performances at temperatures above 35 °C and at high phytoplankton concentrations (Fliermans et al., 1981), and the identification of *Legionella* spp. in hot-water tanks or in thermally polluted rivers emphasizes that water temperature is a crucial factor in its colonization of water distribution systems, its proliferation in the environment, and therefore in infection risk. Thermally altered aquatic environments can result in rapid multiplication of *Legionella* spp., which can translate into human disease. Another important aspect in *Legionella* ecology relates to its nutrient requirements. *Legionella* can proliferate in biofilm, in association with amoebae, protozoa or cyanobacteria (Fields, Benson & Besser, 2002; WHO, 2007). It is commonly found in association with other microorganisms and it has been isolated in cyanobacteria mats, at 45°C over a pH range of 6.9 to 7.6, where it was apparently using algal extracellular products as its carbon and energy sources (Tison et al., 1980).

Naegleria fowleri and *Acanthamoeba* spp., possible emergent waterborne pathogens, can proliferate in warmer water columns.

commonly, the amoebae may be found in bodies of warm freshwater such as lakes, rivers, geothermal (naturally hot) water such as hot springs and drinking-water sources, warm water discharges from industrial plants, poorly maintained and minimally chlorinated or unchlorinated swimming pools, and soil (Behets et al., 2007; Vivesvara et al., 2007; Blair et al., 2008; Jamerson et al., 2008). Infections with *Naegleria* are very rare and occur mainly during July, August and September, usually when it is hot for prolonged periods, causing higher water temperatures and lower water levels. But it is worth noting that infections can increase during heat wave years. *Acanthamoeba* spp. are microscopic, free-living amoebae that are relatively common in the environment. This amoeba has been isolated from water (including natural and treated water in pools or hot tubs), soil, air (in association with cooling towers, and heating, ventilation and air conditioning (HVAC) systems), sewage systems and drinking-water systems (shower heads, taps) (Bartram et al., 2007).[10]

Additionally, *Acanthamoeba* spp. contain several bacterial endosymbionts which can be human pathogens (e.g. *Legionella* spp.), so they are considered to be potential emerging human pathogens (Schmitz-Esser et al., 2008).

Cyanobacteria are ubiquitous autotrophic bacteria. Several species produce different toxins that act with different mechanisms and have been associated with acute human intoxications after exposure through drinking and bathing waters (Funari & Testai, 2008). The most investigated group are the hepatotoxins, which include about 80 different congeners of MCs, and nodularins (NODs), all inducing acute effects in the liver. Environmental factors can also influence cyanotoxin production, but their role is not fully understood: some studies on MCs showed that the variations of parameters such as light, culture age, temperature, pH and nutrients give rise to differences in cellular cyanotoxin content (Sivonen & Jones, 1999). Among these parameters, increasing temperature represents an important factor for increased cyanobacterial proliferation (often associated with increase of nutrients and decreased salinity) and a main factor for their poleward movement and possibly their toxicity. The proliferation of cyanobacterial blooms has been described in the majority of European lakes used for drinking-water supply and recreational purposes, so that this situation has given rise to concern for human health, which can be affected by consumption of contaminated drinking-water or food or by ingesting water during recreational activities. Indeed, some experimental evidence on *Aphanizomenon* spp.-isolated cultured strains have indicated that increasing temperatures up to 28 °C caused an approximate two-fold increase in paralytic shellfish poisoning (PSP) toxin production (Dias, Pereira & Franca, 2002).

Cylindrospermopsis raciborskii, known as a species of tropical origin, has been found to occur with increasing frequency from the mid-1990s in Germany, France, Italy, temperate North American areas, and more recently it has been described as the predominant component of the phytoplankton community of different Portuguese rivers and reservoirs (Saker et al., 2003).

One of the key other water-related health issues is algal blooms. The abundance of toxic HAB-forming[11] dinoflagellates has significantly increased since 1990 in Europe in areas in the North Sea and north east Atlantic (Edwards et al., 2006) and in some marine ecosystems, such as the Grand Banks area of the north west Atlantic (Johns et al., 2003) and Baltic Sea (Wasmund & Uhlig, 2003). This shift in HAB community composition has been linked to increased sea-surface temperature. In the eastern Mediterranean, sudden loads of high nutrient water linked to heavy storms led to an increase in phytoplankton biomass and the dominance of toxic HAB species (Spatharis et al., 2007). Biogeographical boundary shifts in phytoplankton populations made possible by climate change have the potential to lead to the poleward spread of HAB species normally suited to milder waters (Edwards et al., 2006). Norwegian coastal waters of the North Sea have experienced a decrease in salinity related to increased precipitation and increased terrestrial runoff and this has been associated with an increase in abundance of several HAB-forming species (Edwards et al., 2006). The impact of HAB on human health could include more cases of human shellfish and reef-fish poisoning (EEA, 2008), as well as impacting the reliability and operating costs of water systems (Bates et al., 2008).

Major changes in policy and planning are needed if ongoing and future investments in water supply, sanitation and health are not to be wasted. Technologies capable of adapting to the range of climate change scenarios need to be identified and prioritized. Technologies and planning are needed that can be adapted to cope with multiple threats, including but not exclusively limited to the effects of climate change.

---

10   *Acanthamoeba* is capable of causing several infections in humans like *Acanthamoeba* keratitis, a local infection of the eye that can result in permanent visual impairment or blindness (Auger & Lally, 2008).

11   HAB: harmful algal blooms

Chapter 6

WATER SAFETY PLANS: AN APPROACH TO MANAGING RISKS ASSOCIATED WITH EXTREME WEATHER EVENTS

Roger Aertgeerts
WHO Regional Office for Europe

Water safety plans control risks from source to tap, through the different control points in water treatment plants
© WHO/Roger Aertgeerts

Water Safety Plans: an Approach to Managing Risks Associated with Extreme Weather Events

6.1 Key Messages

Water safety plans (WSPs) are introduced by WHO's *Guidelines for drinking-water quality* (2006) as "…a comprehensive risk assessment and risk management approach that encompasses all steps in water supply from catchments to consumer". The aim is clear: "to consistently ensure the safety and acceptability of a drinking-water supply".

WSPs are likely to be an important instrument for water services to extreme weather events, because of the flexible nature of their component steps.

- WSP team offers the possibility of including meteorologists, hydrologists, and geohydrologists from the beginning the risk assessment risk management process.
- The description of the water supply system allows the identification of areas where changes in the short-term weather pattern or the longer-term climate could create physical problems during periods of unusual precipitation or drought.
- During the identification of hazards and assessment of risks, extreme weather events can be treated in the same manner as other threats to the integrity and functioning of the water supply.
- Determination and validation of control measures allows the assessment of the effectiveness of control measures during quality changes of the resource waters, for example higher pathogen loads during rainfall.
- Risk prioritization allows the correct placement of extreme weather in the overall risk assessment, and hence the allocation of sufficient resources to address the issue.
- Improvement plans can be tailored to include the specific challenges posed by extreme weather.

A WSP is an effective framework for assessing and addressing the risks from extreme weather and ensuring the continued functioning of water supply systems.

6.2 Elements of a WSP

6.2.1 WSP team creation and preparatory activities

The WSP team typically constitutes water supply staff and is a multidisciplinary team with knowledge and experience of the water supply system, and direct understanding of the hazards to continuity of supply, quality and usage of the final product. The adequate integration of extreme weather events in a WSP approach requires that the team either includes or closely liaises with stakeholders who have knowledge of geohydrological and meteorological characteristics and inputs to the catchment area. The risk assessment should include historic information on past flooding and drought events, as well as detailed projections for future changes in hydrological patterns. Broad stakeholder involvement in the design and implementation of WSPs is essential in predicting extreme weather events, predicting the hazards and risks, developing appropriate contingency plans, and designing effective contingency planning. Health system components such as hospitals, retirement homes and so on may have special requirements for continuity of supply or quality of water which should be taken into account when designing WSPs.

Communication should also be established at an early stage with the national health system to (a) understand better the vulnerability of key components of the health system (hospitals, dispensaries, first aid stations, hospices…); (b) understand the role the health system can play in mitigating and responding to cases of flooding and drought, and how mutual support can be organized; (c) gather information on people with special needs or with restricted self-help capacity (elderly, immunocompromised people, those needing extended home nursing care).

6.2.2 Description of the water supply system

The WSP team needs to describe the water supply system to support the subsequent hazard identification and risk assessment process. This description should include information on the catchment, through treatment and distribution to the point of consumption.

European water services draw on a wide variety of sources (rivers, lakes, wells, mines…); these resources need to be fully identified and registered, including their geohydrological characteristics. This registration must be done with a sufficient level of detail to allow a vulnerability assessment. For example, it is not sufficient to merely register an aquifer and identify its linkages with the overlying surface water. Its vulnerability to pollution originating in the catchment area also needs to be registered and assessed. Information is also needed on abstraction and water uses, quantity and available resources, and how both quantity and quality may be affected by the impacts of climate change.

In coastal areas, not only the feedwater of the aquifer needs to be registered but also the potential intrusion of seawater.

Although by no means exhaustive, the following list illustrates some of the concerns that the WSP team will face at this point in the process.

Linked water services. are important to provide aid in case a water service is compromised by an extreme weather event. The physical linkage of water services that operate independently from one another may create additional hazards to service quantity and quality, unless such hazards have been identified prior to the installation of an operational connection. It is therefore important to assess the hazards of emergency inter-connections and take all necessary measures to control them well before any emergency necessitates the operationalization of cross-connections. Links are not limited to piped distribution networks. Experience has shown that water services of major conurbations may be called upon, at very short notice, to take over water supply in rural areas where shallow wells became unproductive during periods of extreme drought. In such cases, connections are done by lorry.

Second is a comprehensive source to tap approach. Within a water supply, there may be various actors and stakeholders that are responsible for different elements of the supply, for example source water protection and/or abstraction, water treatment, water distribution and/or storage, and consumer use. It is important to gather information on and understand all the elements of the water supply, regardless of who is responsible. The water supply does not end at the property line and in-house water storage and treatment are important aspects to consider when describing the water supply, and identifying, assessing and mitigating any related risks.

Detailed information on **land use** needs to be gathered, identifying potential hazards relating to agriculture, industry and human settlement, infrastructure and treatment failures, sabotage or man-made disasters and natural events. Bear in mind that extreme weather may change or compound the potential risk of a hazard or hazardous event. For example, a historically contaminated site may present no hazard during routine conditions, but intensive rains may create a hazardous event resulting in contamination of the water supply.

Staff capacity is also important. The development of an integrated description of the system may require knowledge beyond that possessed by existing staff. The WSP team may need to partner with other stakeholders, including the public health and environmental sectors, to gather this information.

6.2.3 Identification of hazards and assessment of risks

The role of the WSP is to ask:

- what can go wrong at what point in the water supply system in terms of hazards and hazardous events; and

- how likely is the hazard/hazardous event to occur and how severe would be the consequences if it did?
These two activities constitute the risk assessment.

Correct implementation of this step of the WSP requires considerable out-of-the-box thinking capacity. For example, information gathering may have been done on the *composition* of distribution pipes in recognition of plumbo-solvency problems, but pressure fluctuations may have remained unrecognized as a specific hazardous event. This would be an issue in water distribution, but perhaps even more so in sanitation and drainage networks.

Hazard identification and risk assessment inside consumer premises also remains a challenge for many water utilities, and consumers and other stakeholders should be engaged as part of the WSP process.

Table 15 (Typical hazards associated with extreme weather events) lists some of the hazards and hazardous events typically associated with extreme weather events.

6.2.4 Determination and validation of control measures and reassessment and prioritization of risks

"Control measures" are activities and processes applied to reduce or mitigate risks. They are determined for each of the identified hazards/hazardous events; missing controls to deal with identified hazardous events need to be documented and addressed.

Control measures must be considered not only for their longer-term average performance, but also in the light of their potential to fail or be ineffective over a short space of time. This is especially important in extreme weather: certain pathogenic organisms and their toxins may, for example, pose a real risk only under conditions of extreme drought.

"Validation" is the process of obtaining evidence of the performance of control measures. Validation is different from operational monitoring of processes which shows that validated controls continue to work effectively. Validation can for example include visual inspection of the catchment area to ensure the absence of cattle, assessment of the underground travel time in river bank filtration, certification of alternative suppliers, testing of alarms for the sudden discontinuation of UV disinfection, and so on.

"Risks" need to be recalculated taking into account the effectiveness of each control. All identified risks should then be prioritized.

6.2.5 Development, implementation and maintenance of an improvement plan

For risks that are inadequately controlled, an update or improvement plan should be developed. This plan should consider short-, medium- and long-term options and implementation to control or mitigate the identified risk. In the case of small supplies or resource restrictions, prioritization based on significant risks to human health and a phased implementation approach may need to be considered. Improvement plans are not necessarily limited to work within the water service's own installations. The WSP approach emphasizes that equal attention is placed on areas outside the direct control of the water supplier, such as the catchment and point of use (consumer). Improvement programmes identified in these areas would require joint action by water services and other stakeholders. Such joint initiatives are usually welcomed by regulators as they are likely to yield, over time, more sustainable results.

The development of upgrade plans offers a clear opportunity to explore in depth the links between water and sanitation systems in one area. Well-known risks exist during extreme weather events, such as flooding which may contaminate the resource waters with untreated wastewater or from wastewater treatment plants (WWTPs). During extreme drought, contamination may occur from the unsafe use of effluents from under-performing installations. These risks can become much more important in the case of flooding of the wastewater treatment, its partial collapse under increased load factors, or the flooding of sludge drying beds. It is important to review any new hazards that could result from implementation of the improvement programme and recalculate the risks, taking into account the new control measures.

6.2.6 Monitoring control measures

"Operational monitoring" includes defining and validating the monitoring of control measures and establishing procedures to demonstrate that the controls continue to work. For example, in many countries of eastern Europe, leaky pipes and unauthorized connections lead to the ingress of contaminated water. Maintenance of system pressure could be a control measure, and installation of pressure gauges throughout the system an appropriate monitoring measure to ensure that the controls work at all times.

Back-up systems are necessary to ensure continuity of supply if monitoring indicates the failure of a control. For example, if chlorination were to fail and flooding of access roads were to prevent new supplies from reaching the production unit, alternative water sources identified earlier could be allowed access to the distribution network. Of course, risks associated with the use of such alternative sources should be identified and addressed during the initial establishment of the WSP.

6.2.7 Verification of the effectiveness of the WSP

Verification involves three activities which, when undertaken together, provide evidence that the WSP is working effectively.

Table 15. Typical hazards associated with extreme weather events

Catchment		Treatment		Distribution	
Hazardous event	Associated hazards and hazardous events to consider	Critical issues to consider in drawing up a WSP	Associated hazards and issues to consider	Hazardous event	Associated hazards and issues to consider
Meteorology: flooding or droughts	Water quality or quantity change	Power supply	Interruption, loss of treatment, distribution compromised.	Pipe burst	Ingress of contamination
Seasonal variations	Colonization of resource waters by opportunistic invader species	Capacity of treatment works	Hydraulic overload of both drinking-water and wastewater treatment plants	Pressure fluctuations	Ingress of contamination
Unconfined aquifer	Water quality subject to unexpected change esp. after long dry spells.	Failure of by-pass facilities (physical collapse or errors in dimensioning)	Inadequate treatment	Intermittent supply	Ingress of contamination
Well/boreholes not watertight	Surface water intrusion	Treatment failure	Untreated water		
Borehole casting corroded or incomplete	Surface water intrusion	Screen / Filter blockage	Inadequate removal of particulate matter		
Flooding	Quality and quantitative sufficiency of raw water, safety of routing floodwaters from critical areas (population, control installations…)	Flooding	Loss or restriction of treatment works		
		Failure of continuous supply of treatment chemicals	Compromised treatment/disinfection		

- **Compliance monitoring** validates effectiveness and monitoring performance against set limits (water quality targets).

- **Internal and external auditing of operational activities** ensures that water quality is within targeted limits and that risks are controlled.

- **Monitoring consumer satisfaction** is important to ensure that the water distributed by the utility will indeed be used. Any consumer complaints about taste, colour or smell should raise concern that drinking-water may not be safe or acceptable to consumers for consumption.

6.2.8 Preparation of management procedures and supporting programmes

Communication programmes need to be set up to instruct people on safety issues. This may involve reassuring consumers on the continued safety of the water when, for example, temporary discolouration or other organoleptic changes occur due to sudden changes of raw water quality, but it may also mean the management of incidents which compromise the microbial quality of the water. Care should be given to preventing consumers from turning to unsafe water sources if problems do not compromise the safety of the distributed water or are of such a nature that they can easily be controlled by simple in-house treatment.

Standard operating procedures (SOPs) need to be prepared for the management of the system under routine conditions. Equally important is the preparation of incident situation operational procedures (ISOPs) as an integral part of a WSP. ISOPs need to describe in detail the steps to be followed in specific "incident" situations where a loss of control in the system may occur. An efficient, regular review and updating cycle is also important, especially as new information and more refined models of expected impacts of climate change are released almost daily.

Notwithstanding all reasonable efforts to make contingency planning as detailed and comprehensive as possible, it is humanly impossible to foresee all events. Unforeseen events/incidents may, indeed in all likelihood will, occur for which there are no corrective actions in place. In this case, a generic emergency plan should be followed. This would have a protocol for situation assessment and identification of situations that require activation of the emergency response plan.

It is important that impacts of past extreme events, or near misses of service disruption by extreme events, be analysed, as they could be an indicator of a likely future emergency. It is also important that such lessons be shared as widely as possible within the water utility community through specialized networks, journals and so on, and that the WSP be reviewed and updated with the results.

Supporting programmes are activities that support the development of people's skills and knowledge, commitment to the WSP approach and capacity to manage systems to deliver safe water. Supporting programmes include: research and development, training and capacity building, equipment calibration programmes, laboratory intercalibration, preventative maintenance, development and installation of customer complaint protocols and their processing, as well as legal training, finance and administrative systems, records management, communications and public education/engagement programmes.

In order to be effective, the WSP needs to be a dynamic and evolving approach that is integrated into the daily management plans and processes of the water supply.

6.2.9 Periodic review

The WSP team should meet periodically to review the overall plan and learn from experiences and new procedures. The review process is critical to the overall implementation of the WSP and provides the basis from which future assessments can be made.

Following an emergency, incident or near miss, risks should be reassessed and may need to be fed into the improvement/upgrade plan.

Also, after substantial capital investment in adapting to climate change, such as the installation of stormwater reservoirs, or the installation of PAC (powdered activated carbon) columns, a review of the WSP is indicated to see if SOPs and ISOPs are still adequate or need to be refined and reviewed.

A WSP should also be reviewed following every emergency, incident or unforeseen event, irrespective of whether new hazards were identified, and to ensure that, if possible, the situation does not recur and to see whether the response was sufficient or could have been handled better. The WSP and improvement plan should be updated as a result.

A post-incident review is always likely to identify areas for improvement; in many cases, the greatest benefit will result when other stakeholders are also included in the review.

6.2.10 IWRM

The advantages of placing water services in the framework of IWRM have been lauded by many authors. Examples of such advantages are: the reduction of negative externalities (for example, cattle grazing in a water catchment area may harm the water quality, but are not usually managed by the water service) that arise from the uncoordinated use of interdependent water and land resources; opportunity costs which arise when production factors are used of low value/benefit costing; and cost savings achieved by widening the range of management options.

A number of countries outside the EU are making progress towards IWRM through activities in the area of watershed management, clarifying institutional roles and responsibilities, increasing stakeholder participation, and anchoring financing mechanisms in the integrated management concept.

Inside the EU, the legislative framework has been created for climate adaptation strategies for water services to fit into an IWRM approach. The main elements of this legal strategy are:

- Directive 2000/60/EC of the European Parliament and the Council of 23 October 2000 (Council of the European Union, 2000), establishing a framework for Community action in the field of water policy – the WFD;

- Directive 2007/60/EC of the European Parliament and of the Council of 23 October 2007 (Council of the European Union, 2007) on the assessment and management of flood risks.

Important progress is being made in the provision of guidance on water resource management and climate change by the Parties to the 1992 Convention on the Protection and Use of Transboundary Watercourses and International Lakes; the guidance documents resulting from these efforts (UNECE, 2009a) could be usefully consulted in conjunction with the present volume.

Finally, it needs to be recalled that health issues are frequently neglected or underestimated in water resource strategies. Health considerations should, however, be incorporated in regional and national stakeholder dialogues on sustainable water resource management, including the use of these resources for water supply and sanitation. Proper screening and scoping exercises will need to be developed to determine the key health issues, and to define the framework in which they are to be considered.

6.3 The Special Case of Small-Scale Water Supply Systems

6.3.1 Importance of small-scale water supply

Small-scale water supplies comprise different types of supplies which may be categorized by two criteria, i.e. the group of people responsible for their administration, management and operation and the group of users of the supply.

- Private or individual wells comprise point sources, such as boreholes, dug wells or springs, potentially piped into the dwelling or yard, which typically serve a single family or a small number of households (e.g. farms, hamlets), and which are operated by the users themselves.

- For community-managed supply, systems are administered and managed in self-responsibility by the community members (e.g. cooperatives) who are also the users of the water. Community-managed water supplies range from simple point sources from which community members collect water and carry it home to more sophisticated systems which may involve treatment, storage and piped distribution into dwellings or yards.

- Public supplies: systems include administered and managed by a distinct public entity (e.g. municipality, water board association) responsible for the provision of drinking-water to the public in a spatially limited area (e.g. small municipality or town).

Small-scale systems are vital to the water supply of significant parts of the population in all countries of the European region. This applies to both permanent residents and transient users (e.g. tourists, guests). Small-scale water supplies usually prevail in rural areas, including individual farms or settlements, hamlets, villages and small towns, or on small islands. Typically they can also be found in vacation or leisure homes, trailer parks or camping grounds. Displaced, mobile, migrant and temporary populations, including occupiers of temporary homes, pilgrims, nomads, seasonal workers or participants of large festivals or fairs may place additional stress on management and operation of small-scale water supplies. Water supplies serving peri-urban areas (i.e. the communities surrounding major towns and cities) are often beyond the reach of municipal services and organized in the same way.

In the European region, approximately 30% of the total population live in rural areas. Access to improved drinking-water sources in countries of the European region varies between 70% and 100%, and in rural areas between 61% and 100%. Of the population in urban areas 1% is without access to improved drinking-water sources, however, in rural areas, this is the case for 6% of the population or about 16 million people (WHO, 2010). More details for EECCA countries, for countries of south-eastern Europe (SEE) and EU Member States are given in Table 16 (Access to improved drinking-water sources in rural areas in the European region).

6.3.2 Challenges in small-scale water supplies

Small-scale water supplies face a number of similar characteristics and challenges. They are related to their regulatory environment, administration, management, operation or available technical, personnel and financial resources. They include – but are not limited to – the aspects listed below. It should be noted, however, that neither every characteristic described below is necessarily relevant to all small-scale water supplies nor are the challenges limited to small-scale water supplies only.

6.3.2.1 Regulations

- Small-scale supplies are often not regulated or regulated differently compared to larger supplies. The supranational legislation of the EU is an example of this. According to the provisions of the DWD, Member States may exempt supplies serving less than 10 m^3 a day or serving fewer than 50 persons from the minimum requirements of the DWD, unless the water is supplied as part of a commercial or public activity. In cases where regulatory requirements for small-scale water supplies exist, enforcement mechanisms tend to be weak or ineffective, amongst other reasons, due to their large numbers and geographical spread.

- Regulations often base required drinking-water quality monitoring frequencies on the size of the population served.

Table 16. Access to improved drinking-water sources in rural areas in the European region

Area	Total population	Proportion of rural population (%)	Access to improved sources	
			Rural population (%)	Total population (%)
European region	889 574 000	30	94	98
EU	494 769 000	26	92	95
EECCA	276 819 000	36	85	92
SEE	56 429 000	45	59	61
Other countries	93 736 000	29	97	99

Source: WHO/UNICEF (2010).

Minimum monitoring requirements for small-scale water supplies are comparatively rare and typically range between 1 and 4 analyses per year; some jurisdictions even exclude private wells from any monitoring requirements. In combination with non-existent or less stringent reporting requirements, in many countries, systematic evidence on the status of drinking-water quality in small-scale water supplies is not readily available.

6.3.2.2 Attention and sense of responsibility

- Experience has shown that small-scale water supplies typically receive less political attention. Managers and operators of small community-managed or public supplies are rarely organized in professional networks or lobby groups that "mouthpiece" their interests. Therefore, financial and political support, both locally and nationally, is harder to leverage, resulting in limited and inconsistent resourcing.

- There is frequently a low level of awareness and knowledge of potential risks from water to health amongst rural populations – as if to say: "My grandpa already drank our local groundwater and never got sick."

- The inaccurate perception of the importance of water supply for public health protection may lead to a lack of sense of responsibility among local decision-makers resulting in comparatively little political priority and thus underresourcing of water supply.

6.3.2.3 Staff and management

- Small-scale water supplies frequently lack personnel with specialized knowledge. Often non-water professionals or undertrained persons operate the supply. In community-managed or public supplies, staff regularly has many tasks within the community or municipality in addition to water supply. Due to the larger geographical spread, and sometimes remoteness and isolation, operators of small-scale water supplies lack easy access to information, expert assistance and technical support; there is also a low level of networking in scientific and professional communities.

- Frequently, there is a lack of awareness and knowledge and application of internationally or nationally recognized "good" managerial and operational practices, including those recommended by the WHO *Guidelines for drinking-water quality* or relevant international standards. Integrated risk assessment and risk management approaches, such as the WHO-recommended WSP approach, are not extensively applied.

6.3.2.4 Water resources and treatment

- Small-scale water supplies are more vulnerable to contamination. In many rural contexts, sanitary protection of drinking-water sources is inadequate; protection zones are often not established. Especially in agricultural areas, common critical pollution risks include cattle breeding and wildlife, poor manure management or inadequate local sanitation practices which frequently result in poor microbial drinking-water quality or elevated nitrate levels.

- The use of water treatment technologies is generally limited and not necessarily consistent with source water quality. In many rural settings, groundwater is used for drinking purposes without disinfection, regardless of its contamination level. Heavy rainfalls and thaw have been reported to pose significant stress to small-scale treatment systems. Small-scale water supplies are also expected to be less resilient to quality and quantity (e.g. water scarcity) issues induced by climate change.

- Small-scale water supplies are more vulnerable to breakdown. Maintenance of infrastructures is often limited due to the lack of adequate resourcing, spare parts or building materials. In consequence, aged supply infrastructures of in principle improved sources are often disrupted or not in working condition. This and the lack of electricity limit operations, frequently leading to intermittant supply with negative impacts on personal, domestic and food hygiene conditions. Users may also turn to other, potentially unimproved and therefore unsafe sources as alternative sources of water supply.

- Small-scale water supplies have relatively greater capital costs of technical installations and also per unit costs of materials and construction are generally larger.

6.3.3 WSPs and small-scale water supplies

The discussion on WSPs done with a focus on reticulated systems in section 6.2 also holds for small-scale water supply systems. Especially in view of the specific challenges of such systems, the introduction of WSPs is likely to lead to more reliable operation, and support proactive supply management and operation with a focus on prevention. Development of a WSP can encourage operators to have a fresh look at their supply system, and to develop a better understanding of the hazards and risks their system faces including in extreme weather conditions that may not yet have been experienced by the current operators. It may lead to a step-wise improvement and upgrading over time, and to a better substantiation of requests for additional resources to the parent authority.

6.4 Water Safety and Bulk Transport of Water in Extreme Weather Conditions

6.4.1 Water supply by tanker during drought conditions

Under conditions of extreme drought, some European countries have been forced to consider and implement shipment of drinking-water through bulk transportation to outlying stricken areas. While for some countries this is an emergency measure, for others it has become a matter of routine.

Transfer of drinking-water is a practice that has been followed at least in the past 30 years in the Greek islands of the Aegean, with regards to the increasing demand for water during the summer period due to tourism influx in combination with the rainfall shortage and the inadequate infrastructure for local water collection.

The first to introduce it was the Navy, that constructed special ships (tankers) to supply the warships with water during their patrols in the Aegean. The only requirement for the water was that its origin could guarantee the adequate quality of the water for drinking purposes. This practice is also followed by the private tankers that carry drinking-water to a considerable number of small islands, every summer. Usually the water source belongs to the water distribution system of a city, of a well tested quality, that is proved for every travel by a microbiological test of the water according to the DWD. Once per year, a test to verify the chemical indicators of the water is mandatory under the Directive. In addition, the material of the water container of the ship during the travel should be of a quality that doesn't allow oxidation. The quality of the water should also be checked microbiologically at the point of the final supplier.

Although the practice looks quite simple, it implies elevated costs for the consumer, as usually the price is nearly 10 times the price of the water from the water distribution system. In comparison with the sea water desalination plants that present a sustainable alternative for drinking-water production, the above practice is very expensive. It is not recommended for repetitive use with regards to the costs. On the other hand, it may appear attractive, for isolated and non-repetitive cases that demand huge quantities of water for a short period of time, as was the case in the summer of 2009 for Barcelona and Cyprus. Detailed analytical protocols are available from the WHO Project Office at the coordinating unit for the Mediterranean action plan under the Barcelona Convention.[12]

6.4.2 Elements of technical guidance for bulk drinking-water transport under drought conditions

At the time of writing, WHO is in the process of developing specific guidelines for bulk transportation of drinking-water, to be included in the next revision of the *Guidelines for drinking-water quality*.[13] The following salient points can be noted.

6.4.2.1 System risk assessment

A system risk assessment that considers (1) water source, (2) container materials and design, (3) best sanitary practices for filling, transport, storage and delivery and (4) monitoring and reporting helps to ensure acceptable delivered water quality.

Bulk provisioning for drinking-water should be provided by a certified and well-operated drinking-water treatment plant, distribution systems or other municipal water supply sources meeting WHO's or the user country's standards and needs. A hierarchy for desirability of bulk water sources is as follows: treated and disinfected water > treated water > protected groundwater sources (springs and wells) > unprotected surface water with minimal human impact > unprotected surface water impacted by anthropogenic contaminants.

Materials used for bulk water containers, pipelines and fittings are important, because materials perform differently under different environmental conditions, for example temperature, soil corrosiveness and so on. Some materials can leach toxic or organoleptic chemicals, are less corrosion resistant, are more permeable to external contaminants or are more susceptible to

---

12 For details of the protocols please contact whomed@hol.gr.

13 The draft document and comment pages are available on the WHO web site (http://www.who.int/water_sanitation_health/gdwqrevision/fourth_edition_bulkwater_chapter_ckbphil.pdf, accessed 14 September 2010).

biofilm formation. Ease of cleaning is also influenced by container material. Containers should meet water contact materials requirements of the country. Transportation and storage containers should:

- include backflow and back siphonage prevention, and cross-connection prevention;

- be readily cleanable for repeated usage;

- not have had non-food substances in them (e.g. reused fuel or pesticide containers);

- be appropriately labelled as drinking-water containers (symbols, potable water, size of label);

- not use lead- or cadmium-bearing solder on joints or fittings.

To minimize contamination during filling of bulk water containers or charging of water transmission pipelines, sanitary inspections and maintenance of sanitary conditions for "water filling stations" are necessary. Site conditions at the water source location where approved drinking-water is loaded onto mobile vehicles, attached to pipelines or bulk water containers should have proper drainage, have no contamination sources and be secure with restricted access to authorized personnel. Transportation and water storage time and weather conditions (very hot or cold) can affect water temperature and water quality, contributing to concerns on odour, taste, bacteria regrowth and biological quality of the water.

The organic concentration and/or turbidity level in source water can affect disinfectant demand and therefore compromise pathogen disinfection efficiency during bulk water filling and delivery. Re-disinfection may be necessary if residual disinfectant falls below recommended levels during storage and transport. Different-disinfectants often have varying disinfection effectiveness against different genres of microorganisms.

6.4.2.2 Operational monitoring and management

Monitoring should be routine and frequent enough to ensure that the bulk water system complies with sanitary requirements. Microbial, chemical and physical water quality monitoring are all important. Appropriate faecal indicators/surrogates for monitoring include the coliform group, enterococcus or faecal *Streptococcus*, *Bacteroides*, coliphage and *Clostridia*. Also, specific pathogen surrogate or reference pathogen (e.g. *Cryptosporidium*, enteroviruses) monitoring may be useful when problematic in water supplies. Rapid methods (e.g. quantitative polymerase chain reaction) available for faecal indicators and select pathogens can provide timely information on water quality.

Monitoring to ensure an adequate level of residual disinfectant in the bulk water systems is important to control growth of microorganisms and inhibit biofilm formation. Monitoring should be conducted at least at the water filling or connection point, in the pipeline or bulk water transportation system during storage, transmission and at the point of delivery to the consumer.

In addition to operational monitoring, regular maintenance including scheduled inspection, cleaning, repair and asset replacement are key to ensure good water quality within the bulk water systems. Inspections should check for leaks, non-approved paint/coatings on surfaces, chips, damage and degraded gaskets. Repair and cleaning activities can pose contamination risks. Biofilms, rust and sediments may present different cleaning challenges. Transit and storage distance, system retention time and conditions of pipelines and containers are other key factors that influence water quality as well as container integrity.

6.5 General WSP Checklist

- Functioning WSP team is in place.

- Methodology by which a WSP is developed and agreed is in place.

- Ongoing commitment and resource support is confirmed by senior management of water supply organization.

- Water quality targets are identified and used as benchmarks to verify adequacy of WSP.

- The water supply system is accurately described from the catchment, through treatment and distribution, to the consumers' point of use.
- Stakeholders directly or indirectly influenced or affected by water safety are identified and engaged by WSP team.

- Hazards and hazardous events affecting the safety of a water supply are identified (based on local knowledge, visual inspections, historical data, and predictive information).
- Risk presented by each hazard and hazardous event is assessed and prioritized.

- Controls or barriers are established or confirmed for each significant risk, with effectiveness validated.

- Shorter- and longer-term improvement plans are developed.

- Operational monitoring is continuously carried out along with associated corrective actions when operational targets are not met.
- WSP is verified as working effectively through compliance monitoring, including end-point testing and auditing.

- Accurate records are kept, including management procedures maintained for transparency and justification of outcomes.
- Programmes to support WSP implementation are executed or planned (e.g. training programmes, calibration of equipment).

- WSP is regularly reviewed, including hazards, risks and controls.

CHAPTER 7

ADAPTATION MEASURES FOR WATER SUPPLY UTILITIES IN EXTREME WEATHER EVENTS
Jim Foster
Drinking-water regulator, United Kingdom

Flooded drinking-water treatment plant, Mythe, Gloucestershire (United Kingdom), 2007
© British Geological Survey

Adaptation Measures for Water Supply Utilities in Extreme Weather Events

7.1 Key Messages

Extreme weather events could affect the efficiency of drinking-water treatment processes and the stability of drinking-water in distribution. Water suppliers can prepare to minimize the impact of extreme events on the service rendered in a number of ways.

General

- Strengthen ongoing communication with meteorological forecasting offices.
- Implement pro-active measures to identify changes in quantity and quality of the resources water.
- Identify alternative resources and ensure their timely use.
- Plan in advance measures that should be taken if a critical site becomes unavailable due to an extreme event.
- Develop emergency plans including the role of each organizational component, individual or team that will respond.

Adaptation measures for drought

- Review plant and equipment to ensure it remains appropriate for any reduction in flow or change in source quality.
- Reduce leakage in a pro-active manner by re-evaluating the acceptable economic level of leakage admitted under nominal conditions.
- Where pressure or flow reductions are implemented, take care to ensure a minimum supply as well as to meet the specific needs of vulnerable groups.
- Restrictions on water use need to be carefully communicated to consumers.

Adaptation measures for floods

- Review the siting of the water treatment plant in the floodplain.
- Develop in a pro-active manner site-specific plans that identify not only safe actions and escape routes for staff, but also minimize the impact of floodwater on operational equipment.
- Develop operational programmes to regain drinking-water supply systems after flooding.

Emergency distribution of alternative water supplies

- Make plans to meet the drinking and sanitary requirements of the population at all times.

7.2 Vulnerability of the Water Cycle to Extreme Weather Events

Climate change and extreme weather events will affect many aspects of the water supply infrastructure, as described in Chapter 2. In relation to drinking-water supplies, this will particularly impact on the availability and quality of raw water, which in turn could affect the efficiency of drinking-water treatment processes and the stability of drinking-water in distribution. In general terms water suppliers will need to consider adaptations to deal with greater variation in water quantity (both the availability of raw sources and the demand for suppliers); and greater variability in raw water quality and resulting requirements to ensure the supply of safe drinking-water. In order to ensure a safe and continuous water supply during extreme weather events measures must be taken across all aspects of the water supply system – with regard to water sources (catchments and aquifers); water collection, treatment and distribution, and also in the management of demand and water use on premises.

Particular issues that may affect drinking-water supply systems include the following:

- an increase in the intensity, severity and frequency of extreme weather events;

- reduced availability of water in rivers, reservoirs and aquifers, which also means lower quality in some cases due to reduced dilution of pollutants;

- different treatment of water supplies due to lower quality of water in the environment, which will cost more money and use more energy;

- effects on existing sewerage systems, which were not designed to take climate change into account; more intense rainfall is likely to exceed the capacity of parts of the network and cause local flooding and deterioration of sources of water;

- water quality problems caused by runoff taking nutrients and pesticides from agricultural land, for example, and transferring them into rivers and lakes;

- effects on the structure and operation of dams and reservoirs, for example from increased siltation and slippage of reservoir walls into the water, contaminating it;

- piped systems for both drinking-water supply and sewerage becoming more prone to cracking as climate change leads to greater soil movement as a consequence of wetting and drying cycles;

- increased risk to assets on the coast or in flood plains from flooding, storm damage, coastal erosion and a rise in sea level;

- discolouration and odour problems resulting from higher temperatures and more intense rainfall events;

- likely increased demand for water, particularly at times of reduced availability, exacerbating supply issues;

- financial and economic impacts as well as environmental and social consequences.

Fig. 10 (Schematic illustration of a river catchment) shows the position of water supply and sanitation in a river catchment, with flooding risks arising from river catchment flooding or storm-driven flood waves.

Any kind of extreme weather event has a potential impact on water supply and sanitation, whether affecting individual features or the overall system (see Table 17).

7.3 Adaptation Measures for Drought Events

Drought is a function of water scarcity. Water scarcity may prevail over long time periods and the distinction between operational water management during times of water scarcity and emergency activity in a drought situation is much less clear than say in the event of a major flooding event. Thus the continued supply of safe drinking-water during drought conditions must normally be considered as part of a continuum of operational management during times of water scarcity. Many adaptation measures adopted in water-scarce regions are equally relevant during periods of drought in either water-scarce or other regions.

In drought events, it is vitally important to consider the potential impacts (and thus adaptation options) across the whole water supply system, from source to tap. The interrelation of measures across the supply system must also be assessed when considering adaptation measures.

7.3.1 Adaptation measures in advance of an extreme event – drought

7.3.1.1 Source and reservoir management

The careful management of water resources is fundamental to the supply of adequate safe drinking-water during drought conditions. Water suppliers must work with a variety of stakeholders to understand the climatic and meteorological conditions in which they are operating and also with those responsible for environmental protection and the management of water and land-use within catchment areas.

As part of their comprehensive risk assessment and management activities incorporated in a WSP, water suppliers need to identify risks to the availability and quality of resources posed by drought scenarios.

In particular water suppliers need to have in place standing agreements for communicating with meteorological forecasting units to ensure that long- and short-term forecasts of dry periods can be highlighted and to identify trigger points at which prepared drought management plans can be activated.

Suppliers should work with meteorological and environmental agencies to agree statistical estimates of models of a variety of scenarios. These should typically be based on drought "return periods" or frequencies. Thus scenario details can be created for situations that might be experienced annually, once in every five years, or once every 20 years, for example. Where appropriate, these scenarios can be coordinated with other water resource management activities such as river basin management plans prepared under the European WFD. Suppliers should then use these scenarios as planning assumptions to assess the risks to water supply and quantity.

The primary impact of drought or water scarcity on drinking-water supplies is one of resources or availability. Where possible, adaptation measures to assist in the management of drinking-water supplies during such periods should be put in place in anticipation of future drought conditions. These measures are also likely to form part of long-term water resources planning and may also contribute to several complementary objectives of the water supplier (e.g. planning for population growth). Table 18 (Examples of adaptation measures) offers some examples of proactive adaptation steps.

In some regions of extreme water scarcity roof-top rainwater harvesting offers a reliable source of water, and a minimum quality for safe drinking-water can be achieved with moderate effort. Roof-top systems collect and store rainwater from the roofs of houses or large buildings, greenhouses, courtyards and similar impermeable surfaces, including roads. Most of the rain can be collected and stored. How the harvested water will be used depends on the type of surface used and its cleanliness, as well as users' needs. Modern roofing materials and gutters, for example, allow the collection of clean water suitable for drinking and other domestic uses with limited treatment, especially in rural areas without tap water.

Water quality and safety may also be adversely impacted by drought/water scarcity. These are some of the probable primary impacts.

- There may be a general decrease in raw water quality due to less dilution in source waters.

- Surface water sources may vary greatly due to reduced flow patterns, draw-down of storage reservoirs, and changes in reservoir limnology.

- Groundwater may be more susceptible to contamination than previously due to changes in hydrogeological flow patterns produced by a changing water table level and the associated impact on saturated soil layers.

Fig. 10. Schematic illustration of a river catchment

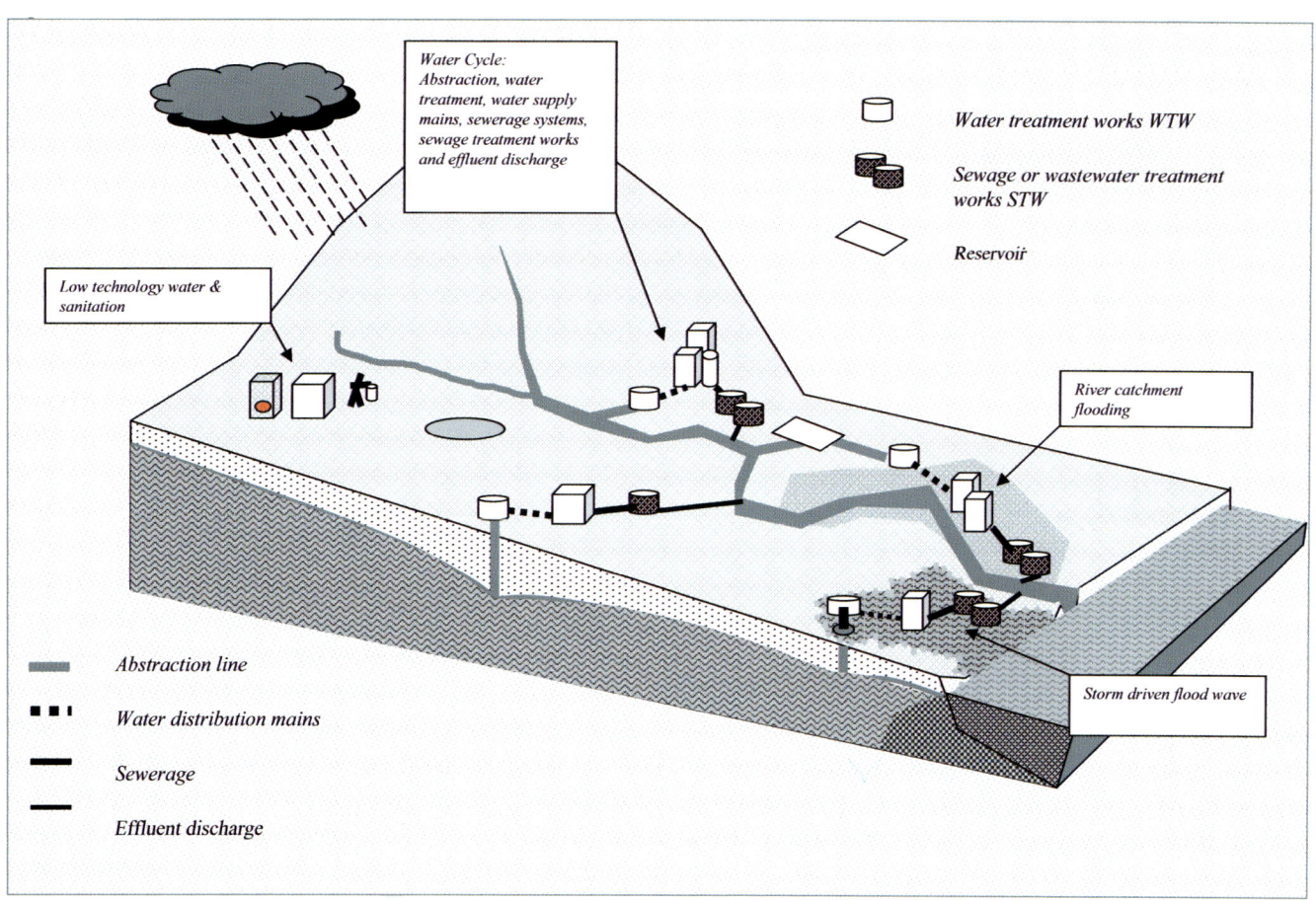

Source: Centre for Environmental Health Engineering, Faculty of Engineering and Physical Sciences, University of Surrey, England.

Table 17. Potential impact on feature or system in water supply and sanitation

Extreme event	Potential impact on feature or system										
	Source water	Abstraction system	WTW	Water supply network	Usage	Sewerage network	STW	Receiving water	Groundwater	Ecosystems	Other
Severe drought	- Reduced quantities available (H) - Reduced quality of surface water sources available (H)	- Reduced water levels at abstraction point adversely affect abstraction rate (M)	- Influent deterioration causes product quality reduction (M) - Reduced throughput affects performance (H)	- Pressure reductions increase infiltration risk (M) - Quality reduction from low flow long residence time in mains (H)	- Increase in demand (H) - Possible rationing of demand (H)	- Low water use causes sewage to break down in foul sewers (M) - Solids deposited in combined sewers (H) - Sediments in surface water sewers harden (H)	- Influent quality adversely affects treatment (H)	- Effluents from STW reduce water quality (H)	- Depletion of ground-water aquifer (H)	- Grey water recycling minimizes potable water supply/demand for irrigation. etc. (H)	- Shallow well systems run dry (H) - Food irrigation needs in rural areas lead to increased demand (H)
Prolonged and extremely high ambient temperatures	- Raw water temperature rise results in lower DO (H) - Water quality more likely to be worsened by upstream STW effluent (H)	- Abstraction system adversely affected (L)	- Lower DO adversely affects bio-water treatment systems, i.e. slow sand filters (H) - Associated operational equipment failure (L)	- Operational problems resulting from temperature effects (L) - Increased water temperatures adversely affect bio-water treatment (H)	- Potential for substantial increase in demand (H)	- Associated operational equipment failure (L) - Impact on surface water, combined and foul sewer networks (N)	- Lower DO adversely affects treatment (L) - Significant variations in bio-process performance, i.e. fixed film processes (M)	- Existing water quality more likely to be worsened by STW effluent, low DO, etc. (H)	- Associated operational equipment failure (I) - Impact on water quality and quantity (L)	- Increase in potable supply demand increases grey water quantities (H)	- Food crops exert increased irrigation demands (H)

Table 17. Continued

Extreme event	Potential impact on feature or system										
	Source water	Abstraction system	WTW	Water supply network	Usage	Sewerage network	STW	Receiving water	Groundwater	Ecosystems	Other
Extensive river catchment flooding	- Water quality deterioration (H)	- Associated operational equipment failure (L) - Intake system flooded (H)	- Flooding of essential unit process (H) - Process performance adversely affected by poor raw water quality (H)	- Pumping stations flooded (H) - Service reservoirs polluted (L) - Flooded taps & float valves allow contamination (M)	- Increased demand for emergency supplies from areas adjacent to flooding (L)	- Solids deposited in inundated surface water sewers (M) - Foul water & combined sewers overflowing, floodwaters contaminated (H)	- Flooded STWs contaminate floodwater (H) - Diluted influent adversely affects treatment (H)	- Flooded STWs & sewer overflows contaminate surface water (H) - Water quality deterioration (H)	- Borehole pumping control and treatment installation failure within flooded area (H)	- Black water treatment processes widely spread, some more likely to be vulnerable to local flooding with associated pathogen risks (H)	- Local low-technology systems very badly affected by flooding (H)
Extreme storm event flooding	- Surface water sewer outfalls contaminate local surface waters (L) - Combined sewer stormwater overflows contaminate local surface waters (M)	- Abstraction system adversely affected (L)	- Local flooding causes short term WTW failure (M)	- Associated operational problems resulting from local flooding (L) - Physicochemical and bio-systems both affected by influent quality variation (H)	- Likelihood of impact on local usage patterns (L)	- Local overload leading to surface water & combined sewer surcharge and flooding (H)	- Local STW system adversely affected by severe overload (H) - Associated local STW operational problems or short-term system failure (H)	- Associated short-term water quality deterioration (L)	- Associated short-term water quality deterioration (L)	- Black water treatment processes widely spread, some more likely to be vulnerable to local flooding with associated pathogen risks (H)	- Local low-technology systems adversely affected by flooding (H)
Extreme storm winds	- High seas and flood waves adversely affect estuary water quality (M) - Storm-driven sea flooding causes saline intrusion into wells and local surface waters (M)	- Operational failure due to storm damage (L)	- Operational failure of WTW due to storm damage (M)	- Operational failure due to storm damage to service reservoir towers, etc. (L)	- Impact on local usage patterns (L)	- Impact on local sewerage system infrastructure (L)	- Impact on local STW infrastructure (M) - Wind / water surface interaction compromises settling processes (M)	- Increased mixing and DO levels (H)	- Operational failure due to storm damage (L) - Storm-driven sea flooding causes saline intrusion into wells and local groundwater (M)	- Impact on local sewerage system infrastructure (L)	- Storm damage to low-technology systems with poor build quality (H)

Table 17. Continued

Extreme event	Source water	Abstraction system	WTW	Water supply network	Usage	Sewerage network	STW	Receiving water	Groundwater	Ecosystems	Other
Extreme and prolonged cold periods	- Freezing affects downstream base flows and local availability (M)	- Operational failure due to freezing (M)	- Failure of outdoor bio-treatment systems, i.e. slow sand filters (H) - Frozen tanks and open water surfaces (H)	- Lower water temperatures cause contraction-related mains failures (M) - Freezing in service reservoirs affects supplies (M)	- Impact on usage patterns (N)	- Associated operational problems in foul water pumping stations, etc. (M) - Surface water drains and inlets unable to accept melt water (H)	- Freezing adversely affects fixed film treatment processes (H) - Associated operational problems in sedimentation tanks, etc. (M)	- Frozen receiving water, short-term local water-quality deterioration caused by effluent concentration (L)	- Aquifer recharge rates adversely affected (H)	- Adverse impact on grey water treatment processes, i.e. reed beds (H)	- Frozen wells and open water surfaces (H) - Low-technology water-flushed sanitation systems adversely affected (M)
Extreme and prolonged snowfall periods	- Snow melt waters adversely affect quality (M) - Snow melt waters cause local flooding (M)	- Operational failure due to snowfall depth (L)	- Operational failure of WTW due to snowfall depth (L)	- Adverse impact on service reservoirs (L) - Impact on maintenance programmes lead to local failures (L)	- Impact on local usage patterns (L)	- Adverse impact on local sewerage systems (L)	- Adverse impact on fixed film treatment processes (L)	- Impact on receiving water (N)	- Impact on aquifer recharge (L)	- Adverse impact on grey water treatment processes, i.e. reed beds (H)	- Snowfall damage to low-technology systems with poor build quality (M)

Source: Centre for Environmental Health Engineering, Faculty of Engineering and Physical Sciences, University of Surrey, United Kingdom.

Notes: Level of risk: high (H), medium (M), low (L), insignificant (I); DO: dissolved oxygen.

A distinction is made between "extensive river catchment flooding" and "extreme storm event flooding". "Extensive river catchment flooding" is primarily related to excessive snow melt water overloading the river system, or after long periods of heavy rain. "Extreme storm event flooding" relates to a single storm event of extremely high intensity, and flooding is likely to be associated with the overloading of the surface water or combined sewerage systems. Impacts are generally experienced as the rain falls for only a short time after the storm has passed.

Table 18. Examples of adaptation measures

	Adaptation activity	Examples of adaptation measures
Water quantity/resources	Strategic water resource planning (25 years +)	• Interconnection of reservoirs between wet/dry areas (inter-regional transfers) • Variety of sources – might need WTW process capable of dealing with river source at one time of year and stored (reservoir) source at another • Silting of ponds, reservoirs, collection chambers and intake works (due to land degradation and increased erosion) • Improving infrastructure resilience • Resource optimization by supervision systems (telemetry monitoring systems coupled with Supervisory Control and Data Aquisition (SCADA) and automated control)
	Alternative sources	• Standby sources that are operated regularly to ensure they work when required, and sampled so operators know quality • Small suppliers with no local alternatives • New water sources – risk assessment: reclaimed water, desalination, etc.

CASE STUDY 5 : ROOF-TOP RAINWATER HARVESTING IN A SEMI-ARID CLIMATE

A pilot project was launched in a small village in central Anatolia in Turkey – a semi-arid climate – to create awareness about water harvesting technology and to develop a model for replication. The intervention was implemented at 30 houses in the first phase of the project, selected together with the community.

Now the pilot site residents enjoy a continuous flow of clean water in their kitchen sinks for the first time in months. In addition to providing clean drinking water in their homes, the rainwater harvesting project has allowed these families to remain in their village.

Source: compiled by B Benli, UNDP Office, Turkey.

CASE STUDY 6 : IMPACT OF CLIMATE CHANGE ON WATER RESOURCES IN AZERBAIJAN

Vulnerability assessment in Azerbaijan suggests a 15–20% decrease in available water resources. Adaptation measures put in place include:
- construction of new reservoirs, and increasing the efficiency of existing ones (US$305 million)
- improvement of water management systems (US$12 million)
- reconstruction of existing water and irrigation systems (costs to be confirmed)
- reducing demand through the use of water-saving technologies (US$ 418 million)
- afforestation (US$ 10 million).

These adaptation measures are expected to save 10 billion cubic metres of water to help deal with water scarcity and extreme weather events.

Source: compiled by representatives of the Ministry of Ecology and Natural Resources, Azerbaijan.

- An increase in raw and treated water temperatures.
- Increased temperature and nutrient concentrations in surface waters that may also increase the likelihood and extent of algal blooms (and associated cyanobacteria).

Hydrological extreme events such as droughts and floods, surface runoff, increasing solar UV radiation, temperatures and evaporation, are some of the outcomes of climate change which may seriously impact surface water reservoirs used for drinking-water supplies. Indeed, changes in water basin levels, concentrations of nutrients, water chemistry and the growth of toxic phytoplankton directly affect the quantity and quality of raw water as well as treatment practices required for the production of drinking-water.

Adaptation measures in themselves may also impact on the safety of water supplies, for example through the need to mix waters with different chemical matrices, or the potential to introduce novel species if large-scale raw water resource transfers are made.

Water suppliers should review the risks to water quality in catchment areas that may result from the above impacts as changes to risk mitigation/control measures may be required.

As with adaptation for water resource yields, where possible measures to assist in minimizing the impact of variations in the quality of source waters should be put in place in anticipation of future drought conditions. Table 19 (Examples of proactive measures) gives some examples of what can be done.

Table 19. Examples of proactive measures

	Adaptation activity	Example of adaptation measures
Source water quality	Enhanced monitoring to detect deterioration in drought (or when drought conditions are predicted)	Enhanced monitoring of: • turbidity/physical quality • indicator organisms (pathogen loading) • algal species and counts • broad chemical screens (e.g. GC/MS scan) for emerging contaminants • limnology – risks of low draw-down, storage reservoir inversion (use of bubblers, forced currents, etc.) • vector-borne diseases (open reservoir management) • emerging risks – suggested chemicals/pathogens, viruses, etc.
New sources	Use of alternative/standby sources	• Knowledge of quality of sources • Pre-event trials/experimentation into impact of mixing water matrices

CASE STUDY 7 : UNPRECEDENTED CYANOBACTERIAL BLOOM AND MC PRODUCTION IN A DRINKING-WATER RESERVOIR IN THE SOUTH OF ITALY

An extraordinary bloom of cyanobacterium *Planktothrix rubescens* was observed in early 2009 in the Occhito basin, a 13 km^2 wide artificial reservoir with a storage capacity of over 270 million cubic metres of water. Maximum algal density exceeded 150 million cells/litre and associated microcystin production occurred in raw water used for production of water for human consumption in surrounding municipalities (serving about 800 000 inhabitants).
Response actions implemented in the first six months were mainly focused on mitigating the risk of toxin presence in distributed drinking-water and efficiently communicating risk information to the target population and authorities.

These included:

a) identification and quantification of microcystins in raw, treated and distributed water, showing:
• uncommon, changeable toxin production in raw water samples with dimethyl-MCRR isomers as the main cyanobacterial metabolites (range 5.0–30.5 µg l^{-1}) also together with MC-RR and MC-LR;
• trace of microcystins sporadically detected in distributed water, always below WHO guideline value;

b) specific treatments using granulated activated carbon (GAC) combined with pre-existing treatment practices, that is, pre-oxidation, flocculation, sand filtration and post-disinfection;

c) management of the drinking-water supply system in order to reduce the contribution of water from the treatment plant down to about 1 100 litres/second by diluting with water obtained from different sources;

d) risk communication on different media channels, including press communications and a dedicated web site within the Puglia Region Portal.

These activities were integrated with a limnological study of the lake to collect information on the nutrients and algal distribution in the water body during the mixing periods, and the thermal stratification. A FluoroProbe was employed to determine the various algae classes (blue-green algae/cyanobacteria, green algae, diatoms/dinoflagellates/chrysophytae and cryptophytae) and to determine the total chlorophyll in the water. The analyses of dissolved and total nutrients were extended to the main tributaries to evaluate the nutrient loads. Concentrations of macro-constituents both in lake and tributaries were assessed to characterize the matrix of dissolved solutes. Levels of trace metals, herbicides and pesticides were also measured in
the lake and in the main river waters.
These measures, implemented through intensive cooperation amongst the main stakeholders, were effective in successfully managing the health risk for the affected populations without requiring any limitation of water uses.
Direct costs relating to the treatment plants can be estimated at about € 700 000 for the first emergency actions, which consisted of the removal of a 0.5 metre-high layer of sand from the standard filtration systems and its replacement with a 0.5-metre layer of GAC (about 400 tonnes). Long-term preventative/remediation actions requires massive investment of about € 10 million to implement a specially designed GAC filtration system to remove trihalomethanes (THMs) and microcystins.
Risk management plans for the medium-long term period were implemented in line with the WHO water safety plan approach and involve specific investment for a new flexible treatment system, and investigation of environmental parameters inducing/affecting/regulating bloom formation in the basin related to seasonal changes, as well as specific training activities for local environmental and health authorities.

Source: compiled by L Lucentini, Higher Institute of Public Health, Italy

7.3.1.2 Water treatment works

In order to consider the impact of drought or water scarcity conditions on the operation of water treatment works (WTWs), suppliers need to consider the original purpose of the works and any treatment processes and the operating performance (both as originally designed and current). Only then can they determine whether or not adaptation measures are required. For example, suppliers need to review plant and equipment (e.g. chemical dosing equipment) to ensure it remains appropriate for any reduction in flow or change in source water quality. Table 20 (WTW adaptation) suggests some issues to consider.

7.3.1.3 Distribution systems

Variations in the quality and quantity of treated water entering a drinking-water supply distribution system (of whatever size and scale) can introduce new challenges to the operation of that system.

The first step that any water supplier should take to manage the provision of water supplies during an extreme event is the pro-active reduction of leakage from the distribution system. In extreme events, particularly droughts, water resources may well be limited and thus it is important to ensure that as much of the available water reaches consumers as possible. Water suppliers will often determine an acceptable economic level of leakage in conjunction with regulators/local government for example a level of leakage at which it is no longer economically viable to make further reductions. However, the assumptions underlying these calculations will need to be reviewed in planning for extreme events where resources will be under greater pressure.

Further possible pro-active adaptation measures for water distribution systems are listed in Table 21 (Adaptation of distribution systems).

Table 20. WTWs adaptation

Monitoring arrangements	Can monitors detect quality changes which may be more severe or happen more quickly than in "normal operation"? Can monitors detect new risks that might occur due to changes in raw water flows and quality?
Flow controls (weirs, pipes, pumps, etc.)	Will these still operate at (significantly) reduced flows? Will these operate under variable flow conditions?
Water losses during treatment	Minimize losses on site through: • audit of all water losses: run sample taps, monitor sample waste, etc. and minimize where possible without compromising quality • optimization of treatment processes: e.g. optimize filter washing regime to ensure minimal water use but maintain adequate backwash/bed expansion (without compromising water quality)
Delivery of treatment chemicals	What impact does the extreme event have on delivery timescales, quantity, etc.?
Storage of treatment chemicals	Is this affected by the extreme event? • Are the chemicals themselves affected by the event (air temperature, humidity, etc.)? What alternative storage options are there?
Chemical dosing equipment	What are the critical control parameters for chemical treatment processes in use (pH, temperature, etc.)? • Are variations of these within tolerance levels? • What are the alternatives (e.g. change in chemical change in process, reduction in throughput)?
Disinfection process	Impact of increased water temperature/changed pH on effectiveness of disinfection • How will flow changes affect the effective Ct of disinfection stages?
Power supplies	Is an alternative power supply available in the case of supply restrictions? • Has it been (robustly) tested? • Does using it affect any treatment stages/pumping options?
Staff/Personnel	How might workers (or their access to work) be affected by the extreme event? What are the contingencies?

Table 21. Adaptation of distribution systems

Quantity/volume-related	Interconnection of distribution networks/resilience; knowledge of key cross-connections, how to operate in emergency situation, water quality implications of operation (is advice to consumers needed?)
	Manage diurnal flows and levels in service reservoirs – may be necessary to retain water in the reservoirs to meet demand in peak hours (night versus day)
	Storage tank/service reservoir flow rate monitoring in order to promptly make interventions to reduce water lost through leakages
Quality-related	Assess need for enhanced monitoring of distribution systems and consumer supplies for physical parameters, chemical parameters (if risk identified) indicator organisms, etc.
	If flows are restricted then need to consider risks of stagnation in tanks and pipes, minimize "dead ends" in the distribution system in advance to reduce risk of stagnant water
	Change in operating regime may cause re-suspension of solids (iron, etc.) in pipework. This may not be a health issue, but may cause mains water to be rejected leading to consumers seeking an alternative (unsafe) source of drinking-water
	Survey of high-risk premises to assess the risk of backflow contamination from premises – localized risk, but need to cover high-risk sites such as industrial buildings, chemical works, sewage works, ports, etc.

CASE STUDY 8 : IMPACT OF WATER SUPPLY AND USAGE IMPROVEMENT, TURKEY

A project in the Saray district of Turkey focused on improved access to safe drinking-water, along with an awareness campaign on efficient water use. The Saray municipality is situated on the Cubuk plateau, near Ankara, and has approximately 15 000 inhabitants. The main water supply pipe of the municipality was 25 years old and in disrepair. The old pipe has fractured and leaked frequently, and in 2006 this caused a loss of 50 000 tonnes of water.

Furthermore, the material of the pipe (asbestos) was not in line with current quality regulations and the water was not treated after it was stored in the main tank. Due to its old age and fragile structure, the pipe fractured and leaked frequently, leaving residents without water and increasing the possibility of contamination.

The Saray municipality, together with the Every Drop Matters regional water partnership project, replaced the ageing pipe with an appropriate-quality ductile pipe, thereby saving significant amounts of water and eliminating the possibility of contamination and other health risks.

In conjunction with the new pipeline, an outreach and awareness campaign centred on the themes of water conservation and efficient use of water was conducted in schools.

Source: compiled by representatives of the General Directorate of State Hydraulic Works, Turkey.

7.3.2 Managing water supplies during extreme events – droughts

7.3.2.1 Demand management

One of the ways to overcome the issue of water scarcity is through demand management. Technological solutions have their own limitations and so technical improvements are often not enough by themselves. However, it is also necessary to change people's perspective and behaviour patterns and the policies of municipalities, so that they use and manage water resources more efficiently. This can take place over a long time period, for example through pro-active consumer education, but often it forms part of an emergency response during an extreme event.

The role of the water suppliers should be central in deciding to reduce water consumption at a local level. This can be achieved either by voluntary measures or by restricting supplies.

Demand may increase during conditions of water scarcity as those consumers with steady consumption trends (e.g. some industrial users) continue to use water, while others (e.g. farmers using irrigation, or households) increase their consumption in an attempt to maintain their way of life in non-drought conditions. In particular, given the associated weather conditions, consumers may increase their outdoor activities and associated water uses (e.g. swimming pools) and use more treated drinking-water to irrigate crops or plants in domestic gardens.

There is a continuum of options, all of which may be implemented through voluntary or mandatory/regulatory means. Fig. 11. (Intervention options in extreme events (floods)) identifies the principal options.

Examples of demand management measures that can be taken are shown in Table 22 (Options for demand management).

Fig. 11. Intervention options in extreme events (floods)

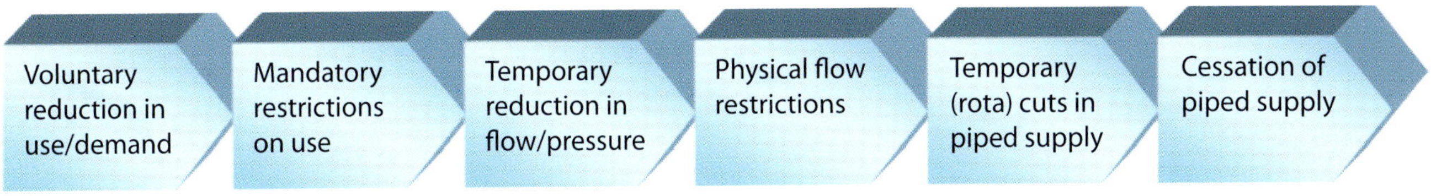

Voluntary reduction in use/demand → Mandatory restrictions on use → Temporary reduction in flow/pressure → Physical flow restrictions → Temporary (rota) cuts in piped supply → Cessation of piped supply

Table 22. Options for demand management

Demand management mechanism	Examples of actions	Issues/risks
Voluntary reductions in use and/or demand	Encouraging increased water efficiency in premises, either through information, short-term incentives, or long-term planning in community building standards. Voluntary reduction in use by large industrial users.	Consumers may not participate in voluntary action, leading to insufficient reduction in demand; localized storage of water may actually increase demand. Impact on economic viability of industry.
Mandatory restrictions on use (e.g. through local or national legislation)	Not permitting certain uses of treated (mains) drinking-water supply, e.g. car washing, garden watering, ornamental pools and fountains, etc.	Confusion over restrictions leading to non-compliance; possible reluctance to participate in future voluntary initiatives.
Temporary reduction in flow/pressure	Reducing pressure and/or flows in supply system on temporary basis at suppliers' sites, e.g. by reducing volumes in storage reservoirs or limited pumping.	Increased risk of ingress/ contamination of supply system.
Physical restrictions to flow/ pressure	Localized flow restricted to certain premises, locations, supply systems (e.g. "trickle flow" device that provides minimum flow for domestic and sanitary uses).	Increased risk of ingress/contamination of supply system. Potential impact on public health due to restrictions in drinking-water/sanitation where consumers do not prioritize these needs.
Temporary (rota) cuts in piped supply	Temporary cessation of supply to specific part of a supply system to enable adequate supplies to other areas. This may be achieved by closing of valves in the distribution system, with the affected areas "rotated" to provide supplies to all areas at some point during any given period.	Increased risk of ingress/contamination of supply system. Potential impact on public health due to restrictions in drinking-water/sanitation where consumers do not prioritize these uses. Encourages consumers to seek alternative (potentially unsafe) supply of drinking-water.
Cessation of piped supply	Piped drinking-water supply is stopped at source or to a specific area.	Increased risk of ingress/contamination of supply system. Impact on public health due to lack of safe drinking-water/sanitation. Encourages consumers to seek alternative (potentially unsafe) supply of drinking-water.

Where pressure or flow reductions are put in place suppliers need to consider the location and needs of vulnerable sectors of the population. For example all premises will need a minimum amount over a certain time period for drinking-water and sanitation. Medical facilities and care homes will need an uninterrupted supply, schools and community buildings may need to be prioritized over other premises, and certain sectors of the population may have specific needs (e.g. for home medical care such as renal dialysis).

If the water supply is not continuous then suppliers will need to carry out a risk assessment of the potential impact on the quality and safety of the supply. This will need to consider the potential contamination risks from ingress (where a pressurized system becomes depressurized) or backflow from contaminated premises in the supply area. Suppliers must ensure that enhanced monitoring of water quality is put in place, in particular for faecal indicators and (where a potential risk is identified) for possible chemical contamination.

Increased health monitoring of the population should also be considered to detect any resulting health/disease burden following changes in drinking-water and/or sanitation arrangements. Water suppliers will need to liaise closely with health professionals to determine the possible risks to consumers and mitigation measures that should be put in place.

Any restrictions on water use will need to be carefully and clearly communicated to consumers so that they are adequately informed of both the resource situation and of actions they can take to assist in resource management. Many examples have shown that where communities are supplied by common or central supplies, then increasing awareness of water scarcity will result in a marked decrease in consumption as individuals take small-scale actions for the collective good. Often suppliers may see demand initially increase in the early stages of any public campaign to raise awareness of water scarcity as people become aware of possible restrictions and perhaps store stocks of water. However this will generally be more than balanced by subsequent reductions in use.

7.3.2.2 Transboundary water resource management and bulk transportation of water

Where highly drought-stressed areas exist it may be necessary to transfer water in from other regions where resources are more plentiful. Transboundary water transfers can take many forms, varying from small-scale intra- or inter-regional transfers (for example between neighbouring water suppliers) to larger scale transfers or bulk transportation.

Suppliers will need to consider the impacts that introducing a different water source into a supply system will have on existing treatment processes, blending and storage arrangements.

7.4 ADAPTATION MEASURES FOR FLOOD EVENTS

7.4.1 Adaptation measures in advance of an extreme event – flooding

7.4.1.1 Proactive adaptation measures: water sources/resources

The primary impacts of floods on water supply systems are too much water in the wrong place (i.e. inundation of the water supply infrastructure), or too great a volume of water of an inappropriate quality for use as a source of drinking-water.

CASE STUDY 9 : TRANSBOUNDARY TRANSFER OF RAW WATER RESOURCES IN AZERBAIJAN

The Kura-Araks river system is a principal source of water for industry, agriculture, residential uses and energy in Armenia, Azerbaijan and Georgia, the Islamic Republic of Iran and Turkey. The rivers are important to regional cooperation as they cross and form many of the borders. Both rivers are seriously degraded in places. Water quality is impaired by the disposal of untreated municipal, industrial, medical and agricultural wastes, and by high sedimentation loads resulting from upstream deforestation. Water quantity is constrained by the use of water for agricultural and hydropower purposes, which also affect the river ecosystem in places.

Integrated, inter-country efforts are under way to evaluate the degree of continuing degradation of these river ecosystems and to take action to halt and reverse damaging trends where necessary. The proposed project aims to ensure that the quality and quantity of the water throughout the Kura-Aras river system meets the short- and long-term needs of the ecosystem and the communities relying upon the rivers. Transboundary cooperation is focusing on:

- fostering regional cooperation
- increasing capacity to address water quality and quantity problems
- demonstrating water quality/quantity improvements
- initiating required policy and legal reforms
- identifying and preparing priority investments, and
- developing sustainable management and financial arrangements.

Activities already begun include:

- developing a water-sharing agreement between Georgia and Azerbaijan;
- establishing sub-basin management councils;
- in each country, implementing at least one on-the-ground investment to address an urgent cross-border water scarcity or pollution conflict;
- carrying out awareness raising, training, seminars and conferences;
- undertaking the creation of data collection, database preparation and information management systems.

Source: Compiled by representatives of the Ministry of Ecology and Natural Resources, Azerbaijan

Where possible adaptation measures should be put in place to assist in the management of drinking-water supplies during such periods in anticipation of extreme events (see Table 23 (Examples of proactive adaptation measures)). These measures are also likely to form part of water suppliers' long-term asset management planning and are similar in nature to those for drought scenarios.

As with planning for drought events, water suppliers must work with a variety of stakeholders to understand the climatic and meteorological conditions in which they are operating and also with those responsible for environmental protection, the management of water and land-use within the catchment area.

As part of their comprehensive risk assessment and management activities under the WSP, water suppliers need to identify risks to the availability and quality of water resources, and the impact of flooding on water supply assets.

Water suppliers need to have in place standing arrangements for communicating with meteorological forecasting units to ensure that long- and short-term forecasts of flooding can be highlighted and should identify both planning assumptions and trigger points at which prepared response plans can be activated.

7.4.1.2 Proactive adaptation measures: water quality

Water quality and safety may also be adversely impacted by flooding events. The primary impacts are likely to be:

- a general decrease in raw water quality due to greater surface runoff and pollution inputs into source waters;
- large scale variations in surface water due to extreme flow patterns, and changes in reservoir limnology;
- groundwater becoming more susceptible to contamination than previously due to changes in hydrogeology;
- an increase in contamination events resulting either from inundation of contaminated land or overflowing of sewerage and drainage systems.

Water suppliers should review the risks to water quality in catchment areas that may result from these impacts as changes to risk mitigation/control measures may be required. As with adaptation for drought events, where possible measures to assist in ensuring the continued safety of drinking-water supplies should be put in place in anticipation of future drought conditions. Example measures are shown in Table 24 (Adaptation activities (floods)).

7.4.1.3 Proactive adaptation measures: treatment and distribution assets (flood protection)

In many cases water supply assets are located close to water bodies or in floodplain areas. In addition to the potential impacts on treatment process and distribution activities that are outlined above on drought (many of which are equally applicable to flood events), water suppliers need to plan in advance the actions that they would take if a critical site (e.g. treatment plant) or infrastructure were to be unavailable due to an extreme event. In the case of flooding, suppliers should work with environmental protection or federal agencies to identify flood risk areas and estimate the extent and depth of flooding that could be expected in a variety of scenarios.

Table 23. Examples of proactive adaptation measures

	Adaptation activity (floods)	Example of adaptation measures (floods)
Water quantity/resources	Strategic water supply system planning	• Interconnection of reservoirs in flood-prone and non-flood-prone areas (inter-regional transfers) • Variety of sources – might need treatment processes capable of dealing with river source at one time of year and stored (reservoir) source at another • Prevention of silting of ponds, reservoirs, collection chambers and intake works (due to land degradation and increased erosion) • Improving infrastructure resilience • Resource optimization by supervision systems (telemetry monitoring systems coupled with SCADA and automated control)
	Alternative sources	• Standby sources that are operated regularly to ensure they work when required, and sampled to provide historical quality data • Small suppliers with no local alternatives • New water sources – risk assessment: reclaimed water, desalination, etc.

Table 24. Adaptation activities (floods)

	Adaptation activity (floods)	Example of adaptation measures (floods)
Source water quality	Enhanced monitoring to detect deterioration in quality associated with peak flows/surges	Enhanced monitoring of • turbidity/physical quality • indicator organisms (pathogen loading) • broad chemical screens (e.g. GC/MS scan) for emerging contaminants • emerging risks–suggest chemicals/pathogens, viruses, etc. • communication links with sanitation operators to proactively share information on inundation of drainage/sewerage systems
New sources	Use of alternative/standby sources	• Knowledge of the quality of sources • Pre-event trials/ experimentation into impact of mixing water matrices
Physical asset protection measures	Flood defences	• Identification of key strategic assets

Fig. 12. Flood protection – order of intervention

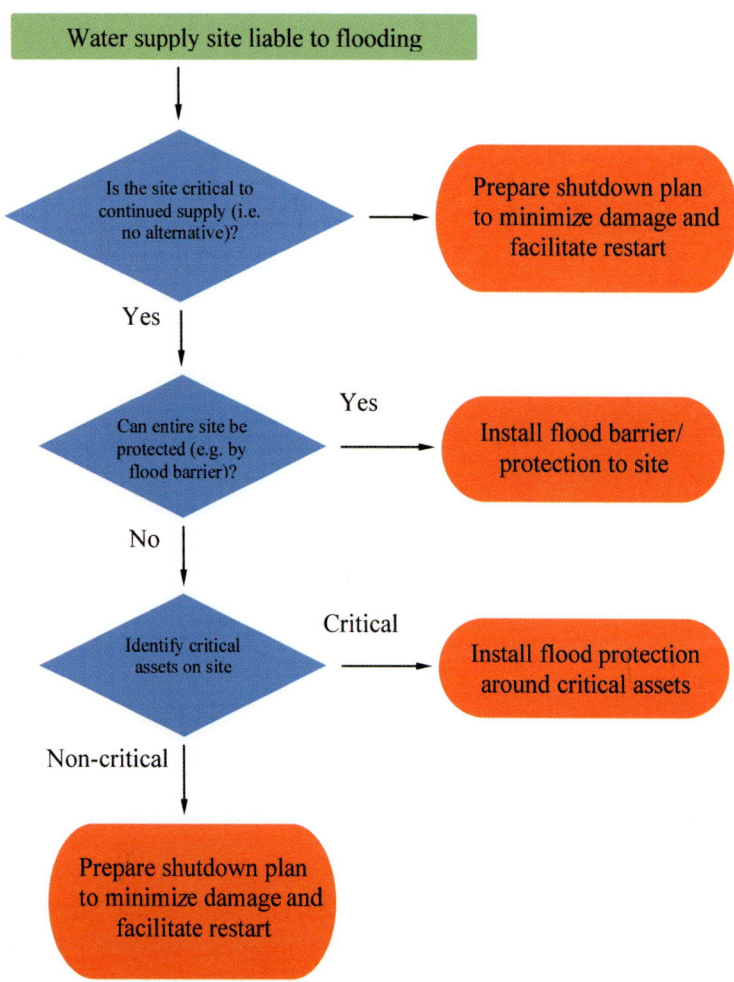

74 Adaptation Measures for Water Supply

Suppliers should have in place a site-specific plan that identifies not only safe actions and escape routes for staff, but also actions that can be taken to minimize the impact of floodwater on operational equipment. The actions taken will depend on how critical the water supply asset is. If there is no option of alternative supplies and continued operation is required, flood protection will be necessary, for example the installation of physical flood barriers to prevent or reduce the impact of rising flood waters. If alternatives are available then actions will be required to minimize damage to ensure that if possible the asset can be restarted after a flood with minimal remedial action and impacts on supplies.

Where flooding of plant or equipment does occur then the risks of operating this equipment should be assessed before use. If electrical systems can be powered down before inundation occurs then damage can be significantly limited. In situations such as this the key action is identifying the electrical components to be dried (either in situ or off-site) or replaced, and once flooding has subsided, or equipment relocated, reinstalled and restarted.

Sites where flooding of electrical components is possible should have simple, easy to understand signs/labels warning operators of the risks of operating flooded equipment. This is especially important for community water supplies, for example WARNING – DO NOT TURN THE PUMP ON. There is a danger of electrical shock and damage to your well or pump if they have been flooded".

The order of intervention is shown in Fig. 12 (Flood protection – order of intervention).

Pre-event adaptations might involve the installation of whole-site flood protection, the fitting of flood protection measures to individual buildings, assets (e.g. intake structures/well heads, power supply buildings) or drainage systems (to prevent surcharging). The scale of flood protection measures will be a balance between the expected severity and likelihood of flooding, and the costs associated with physical protection measures. Typical measures may include:

- where possible making buildings water-tight (e.g. temporary flood gates on doors/openings; enabling closure of external drains (to prevent backflow into the building), and so on;

- increasing flood-resistance of buildings/assets by raising critical equipment and points of potential ingress above the maximum expected flood level;

- installing physical flood barriers (around site or specific asset);

- raising boundary walls at intake sites;

- raising the level of borehole headworks (and ensuring headworks are sealed against surface water ingress).

Suppliers should also note that for alluvial flooding, flood protection measures can often be some distance from the water supply site itself – allowing them, for example, to identify sacrificial floodplain areas to flood in preference to the location of the water supply asset – so there is a need for drinking-water suppliers to work closely with agencies responsible for planning and flood protection on a basin-wide approach.

Flood preparation measures can also involve actions that staff is required to take to minimize damage in the event of a flood, such as checking seals on borehole headworks to prevent ingress, and isolating process streams or electrical equipment. Companies should approach this in a systematic manner to identify risks and how these might be mitigated. Examples of some of the actions that may be taken in advance of a flood are:

- relocation of critical stores;

- relocation of treatment chemical stores;

- closure of unused (non-critical) drainage and overflow valves to reduce risk of backflow;

- isolation of all electrical equipment once flood water reaches a certain trigger level (so as to minimize necessary recovery measures and aid the reinstatement of supplies).

Extreme events, likely to be more common under climate change scenarios, require a rethink of the assumptions made in the preparation of flood adaptation measures. In particular, recent extreme events have shown the need for water suppliers to rethink and review their vulnerability assessments, to rely less on historical forecasts and to plan for more extreme circumstances outside previous assumptions.

7.5 Regaining Drinking-water Supply Systems

7.5.1 Following drought

When replenishing raw water storage reservoirs, reservoir managers should ensure that this is done in a controlled manner, so as not to impact adversely on the quality of water being abstracted for treatment. For example this may involve only allowing the reservoir to replenish at a certain rate to avoid the disturbance of excessive sediment or ensuring adequate dilution of a poor quality source.

Suppliers should also consider the impact on the aquatic system downstream of the reservoir, and the needs of other downstream abstractors, as well as possible environmental/ecological impacts. But priority should always be given to the safety and security of drinking-water supplies.

Suppliers should also be aware that some deterioration in raw water quality may be seen several months or years later as a result of changes during a drought period. For example, colouration of water from upland catchments containing peat soils (due to naturally occurring humic acids) has been seen to increase markedly in summers following a previous drought period. Thus ongoing assessment of risks from catchment areas, informed by local knowledge of catchment hazards, remains important in the post-recovery period.

7.5.2 Following flooding

When regaining drinking-water supplies after a flood it is critical that water suppliers work closely with community leaders and local health professionals/health departments, especially with regard to any precautionary measures that must be taken prior to consumption of water that is supplied (e.g. boiling it before use).

As a general principle, where possible water suppliers should prioritize the use of groundwater/well water where these are well protected (i.e. where they come from a confined or well protected aquifer) in preference to using water taken from rivers or lakes (surface water). This is because the contamination impact on surface water will probably be very much greater. However this will vary with local circumstances and a risk assessment, based on local knowledge of source waters, should be used to prioritize their use for treatment and supply as drinking-water.

Before restarting, flooded treatment works and distribution network disinfection requires a number of planned actions that need to be taken and tailored for centralized, decentralized and community-based production utilities.

Key principles of recovery are summarized in Table 25 (Key principles in recovering a water supply system (summary table)).

7.5.3 Disinfecting and restarting domestic distribution systems (house connections and public buildings)

Water suppliers should prioritize premises where vulnerable consumers may use the water supply (e.g. hospitals, health care centres, etc.), and public buildings to ensure that risks to the general population are reduced. Remediation of domestic premises may be complex and will require clear communication with consumers/householders.

This may involve using a dedicated workforce with knowledge of domestic plumbing arrangements to work alongside the population of an affected area in assessing risks and putting into place actions to remediate the domestic pipe work.

Where the potential for contamination by pathogens is identified by the risk assessment then the domestic system should also be disinfected. The most effective way to do this is one unit, such as a length of pipe, at a time: ensure the supply is not consumed for a set period, introduce a high strength chlorine-based solution into the distribution pipe work (e.g. at the entrance meter or tank), allow to stand for several hours (typically 2.0mg/l chlorine for at least 8 hours) and then flush the system until a low chlorine residual is obtained (typically 0.2–0.5 mg/l). If chemical contamination is suspected

Table 25. Key principles in recovering a water supply system (summary table)

Regaining wells and boreholes	Start with most used critical supply first. **Assess damage:** check pump, check borehole casings/void (using clean steel rod or dip tube). **Rehabilitate:** clear borehole with compressed air, ensure headworks are above ground level and re-sealed, repair pump & ancillary equipment. **Test pump output:** pump out at least 2x borehole volumes, check clarity and basic quality. If not acceptable then re-assess and rehabilitate again (or if repeated, then consider abandoning). Check pump output with required demand/pre-event output. **Disinfect:** ENSURE NOBODY CAN USE DURING DISINFECTION. Check pH is between 6 and 8, and turbidity <5NTU. Add at least 1 litre of 0.2% chlorine solution for every 100 litres of water volume in the borehole and allow to stand for as long as possible. Flush until residual chlorine is <0.5mg/l. **Sample:** sample for indicator bacteria (evidence of pathogens). If possible ensure satisfactory samples from two separate sampling exercises are obtained prior to supply to consumers.
Regaining WTWs	Use skilled, experienced personnel (where possible who have a working knowledge of the particular WTW). Reinstatement priorities should be as follows (complete all prior to supply to consumers): • ensure source protection as far as possible • reinstate physical treatment • reinstate disinfection • reinstate chemical treatment stages • reinstate non-critical treatment (in short term) stages Follow same model as above: **assess/rehabilitate/test/disinfect/sample**

then remediation will depend on the nature of the contaminant but may vary from flushing of pipe work, treatment with specific removal chemicals (and then flushing), through to complete removal and replacement of domestic pipe work. Water suppliers should work with both building owners and health professionals who understand the human toxicological profile of the substance concerned to determine an appropriate remediation strategy.

7.6 Emergency Planning and Institutional Capacity Issues

7.6.1 Emergency planning and preparedness

Water suppliers should ensure that they have a clear understanding of local and regional resilience and emergency response arrangements. These will vary widely and may even change within a supplier's operational area. Suppliers should ensure that they are aware of the key agencies involved, who is responsible for the coordination of response and recovery actions (e.g. area police chief, the local mayor's office, etc.) and what is expected of their own organization. Taking into account appropriate security and confidentiality issues, emergency planners should share information about each other's equipment and capabilities to enable a coordinated response to be planned. This may allow a much larger event to be successfully managed without the need for large stockpiles of equipment or materials. Critical issues to consider are:

- catchments:
  o rethink vulnerability assessments
  o don't rely on history: forecasts are short-term
  o plan for more extreme events – rethink assumptions
  o monitor rainfall levels, flood levels, deterioration in raw water quality

- treatment works:
  o assess local risks and have a plan in place

- distribution:
  o assess risks and mitigation measures, for example access to service reservoirs
  o have controls in place to prevent contamination by "backflow"

- consumers:
  o assess risks in and from buildings in advance
  o understand human behaviour in extreme events
  o prepare clear information and advice.

Suppliers should ensure that formal emergency plans and arrangements are put in place with all agencies and that these are exercised in a variety of conditions. Emergency plans should include key actions to be taken by each organization and contact details for health authorities/ professionals, state, regional and local agencies, key operational control centres (e.g. of ministries/bureaus of environment protection departments, etc.), and emergency services. These contact details and associated protocols should be tested on a periodic basis in conjunction with the partner agency.

Emergency exercises should cover both the crisis phase and the recovery phase, as the restoration and reintroduction of assets is critical. Recovery plans should include planning of measures to ensure that the drinking-water supply is safe for consumption (or appropriate advice is given to consumers).

Water supplier emergency plans should contain clear definitions of roles in an incident including a description of the role of each organization and of individuals/teams who will

Case study 10 : Recovering a Water Supply System after Floods, England, 2007

Extreme rainfall following a prolonged wet period led to unprecedented flooding in parts of England and Wales in June and July 2007. This caused significant disruption to essential services – transport, electricity supplies and the provision of water and sanitation services. Floodwater levels were considerably higher than had previously been experienced and in many cases exceeded the levels that had been planned for.

Over 300 sewage treatment works were flooded and 6 water treatment works (WTWs) were shut down due to flooding, including Mythe WTW which is the only source of piped drinking-water to 340 000 consumers in Gloucestershire. Alternative water supplies were provided via mobile tankers, temporary tanks (bowsers) in the streets and bottled water. Full recovery of the piped water supply took 16 days.

Although there were no direct health effects reported as a result of the water supply interruption, the importance of a holistic (water safety plan) approach to managing risks throughout the water supply system was demonstrated.

Existing plans to shut down the WTW during flood events assisted in the restoration of WTW operations; and existing regulations to prevent major contamination of depressurized water mains was effective in reducing risks to consumers when water supplies were reinstated.

However the recovery of the piped water supply was delayed by a lack of understanding of the role and responsibilities of water suppliers on the part of other agencies responding to the incident. In addition, inappropriate advice about water consumption was initially provided to consumers due to a lack of understanding of risk management in water supply operations.

Knowledge and understanding of the roles and responsibilities of responding organizations is therefore highlighted as a key learning point from this incident. Although a flood plan was in place at the site, the events of summer 2007 were more severe than the water supplier had planned for. The need to re-think vulnerability assessments to take account of more extreme events is also highlighted as a key learning point.

Source: compiled by representatives of the Drinking-water Inspectorate, United Kingdom.

contribute to the incident response. If emergency plans rely on bringing in additional resources it needs to be clear before they arrive what they are there to achieve.

7.6.2 Emergency distribution of alternative water supplies

Where the normal water supply arrangements are disrupted by an extreme event, then suppliers and/or agencies need to provide an alternative supply to at least meet the drinking and sanitary requirements of the population.

The amount of water required by consumers will depend on the priority of need being addressed. The diagram below shows the relative volume required per person for different uses. It should be remembered that water of different qualities will be required for different uses, as shown in Fig. 13 "Differentiated water quality requirements".

Where water is supplied via an existing safe piped system, it is not recommended that this system should be used to provide untreated water. The recovery of such a system to a standard where safe drinking-water can once again be delivered will take a substantial period of time, during which public health may be compromised. Equally, consumers who are expected to do something different from normal – for example boiling water before drinking it where they do not normally do this, or using it only for non-consumption purposes – should be informed clearly by the supplier what they must do. Very careful consideration should be given to the method of providing these volumes of water, and the quality requirements of different uses.

Where consumers are normally supplied with their drinking water via a piped system (be it centralized or a community water supply) and this is no longer available (or its use is restricted by an extreme event), then drinking-water must be supplied to the community. An effective way to distribute water in the absence of "normal" piped water supplies is via the use of mobile water tanks or storage units. These can be in various forms:

- road tankers/trailer tanks
- pillow tanks/bags (large plastic water holders placed on the ground or back of a vehicle)
- bowsers/portable tanks (usually towed behind a vehicle to a static location)
- bottled water.

Where possible water suppliers should use tanks and tankers reserved only for drinking-water. They should be constructed of a material that will not be detrimental to the quality of drinking-water held within it. Even where dedicated tankers are used, then these should be disinfected prior to use (see case study 11). Where suitable quality control measures are in place they may be disinfected prior to an event and stored ready for easy dispersal in an emergency. This must include suitable storage conditions (the tank must not be contaminated and satisfactory test samples should be obtained from stored tanks on a regular basis).

Where containers are not dedicated for drinking-water, then they should be cleaned, flushed and disinfected prior to use. The internal surface of the tank, together with internal surfaces of any fixtures or fittings (taps, pipe work, etc.) should be cleaned with clean water and detergent, then flushed with potable water (if possible via a pressurized supply/pipe), ensuring safe disposal of the flushing water.

Fig. 13. Differentiated water quality requirements

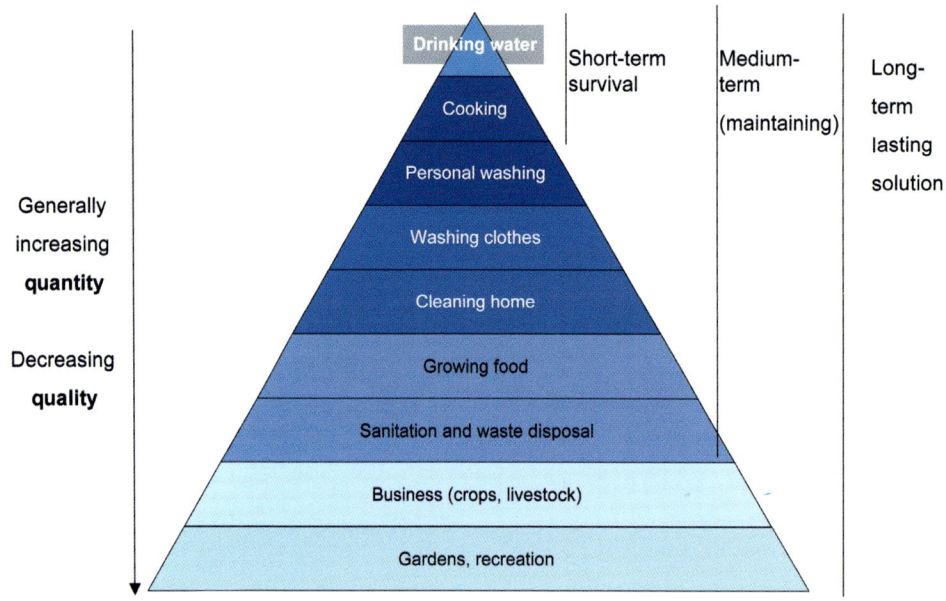

Source: Based on WHO (2005).

Case Study 11 : Water Supply Tanks Disinfection

Alternative water supply tanks should be disinfected internally by the addition of a chlorine solution for 24 hrs (for example 14% sodium hypochlorite solution), and then rinsed (with safe, potable water), and refilled with a safe water supply. After approximately 30 minutes' standing time a sample of the tank contents should be taken to ensure it is safe to supply to consumers. The water should meet all local drinking-water quality requirements applicable to the manner in which it is supplied (and advice given to consumers).

Source: compiled by representatives of the WHO.

Suppliers should ensure that consumers are adequately informed about the purpose of the water supplied and what are appropriate uses and precautions to take. Suppliers should take into account that where consumers have to transport drinking-water from a distribution point or temporary tank/bowser/tanker, then the receptacle used to do this will introduce a risk to safety. Even where the drinking-water stored in the temporary tank is safe, consumers should be advised to boil (and cool) the water before consuming it, to mitigate the risks of contamination during carriage to their premises.

7.6.3 Institutional capacity/Mutual aid

Resources to help in dealing with an extreme event can be shared across a particular sector or group of affected sectors. Water suppliers may put in place "mutual aid" arrangements where collective agreements are made to share a proportion of each other's equipment or resource. For example a water supplier may arrange to share road tankers, static water tanks, pumps, laboratory services or other operational assets from a water supplier who is not affected by an event in their area. It is important that such arrangements are planned for and agreed in advance (based on predictions of what may happen) so that they can be put in place without delay in a major incident. It is good practice to have in place an agreed list of equipment and capabilities that suppliers are willing to share in advance of any event, together with an agreement in principle on the minimum and maximum levels of aid that may be available on a voluntary or commercial basis.

In additional to mutual aid from other water suppliers, aid can be sought and obtained from existing partners and contractors, civil emergency organizations (who may hold strategic equipment stores), national agencies and in many cases the military. Again, it is important to document in advance what aid is available (and how much), where it is located (or delivery lead times), and the processes that need to be followed in order to initiate the provision of this aid (including tried and tested communication arrangements/contact numbers).

7.6.4 Interdependencies/Business continuity

To ensure continuity of a water supply, one needs to be aware of the interdependencies that exists between people, chemical supplies, laboratory facilities, electricity supplies and other factors. A holistic rather than a series of individual approaches is needed to resolve problems. In the event of an extreme event (particularly flooding or adverse weather), access to water supply sites may be affected, as may the provision of services from other suppliers and utility providers.

Water suppliers should ensure that their operational arrangements are extended to consider the impacts of extreme weather events. For example, they may need to consider the issues raised in Table 26 (Impacts and mitigation measures).

Table 26. Impacts and mitigation measures

Impact	Mitigation measure(s)
Access restrictions and unavailability of staff	Shutdown of asset Remote operation Resident staff Increasing chemical stocks and strategic spares
Failure of support services (other utilities, laboratory services, etc.)	Alternative power supplies Increased use of on-site water quality testing

Electrical power supplies are critical to the continued operation of the water supply system. Water suppliers should put in place arrangements with local electricity companies to restore power supplies to key water supply assets as an essential service. It is helpful for suppliers to identify key strategic assets that are critical to the continued supply of treated drinking-water. Service level agreements should be put in place with electricity companies which specify agreed standards of service. Such agreements may include resolutions on the provision of multiple electricity feeds to a site, or a minimum "down time" period, within which mains electricity supplies are restored. Where multiple feeds are in place, a full failure analysis should be conducted to see if there are common points in the electricity distribution system that may cause all independent feeds to fail simultaneously. It may also be appropriate for co-funded projects to be put in place to improve the resilience of power supplies to a particular location or asset. An assessment of the reliability of a power supply should be combined with an assessment of the need for alternative power sources.

Water suppliers should consider the need for either fixed or portable equipment to provide an alternative power supply. Where generators are used, care should be taken to ensure that these are appropriately sized for the site at which they are (or may be) used. This should take into account the continued operation of all key processes, pumps, monitoring and alarms, and also the additional loads placed on the supply by operating changes (for example the starting of pumps). Generators should be regularly maintained and tested to ensure they operate when required. This regular testing should ideally be conducted "on-load" – that is with a typical electrical power demand on the system – to simulate real conditions in the event of a power failure. Consideration should also be given to the mode of changeover between mains power supplies and the alternative source. This may necessitate manual intervention by operators, although various automated changeover options are possible to minimize the impact on the water supply processes in advance of an event occurring.

drinking-water treatment processes and the stability of drinking-water in distribution. More frequent extreme rainfall could lead to increased surface water turbidity and higher numbers of pathogens (and their indicators). This would result in a greater challenge for WTWs, particularly surface water sites. Changes in rainfall may cause more frequent or intense periods of water scarcity and/or drought. This will result in a decrease in available resources and may also increase the chances of resource contamination.

Water suppliers can adapt to prepare for extreme events to minimize their impact on consumers by, for instance, improving the resilience and interconnectivity of supplies, assessing risks to supply systems and treatment processes in advance, and, where possible, putting in place controls to minimize these risks.

7.7 Summary

Climate change is likely to affect the availability and quality of raw water, which in turn could affect the efficiency of

Case study 12 : Water Supply Problems in the Case of Power Cuts Caused by Extreme Weather Conditions, Hungary

In the west Transdanubia region of Hungary a Mediterranean cyclone resulted in a considerable amount of precipitation in the form of snow on 27 January 2009. A great quantity of snow froze on (mainly medium voltage) network cables, which were damaged by its weight and by the accompanying strong wind, thereby causing an extensive power cut. The power cut affected 34 settlements and 89 000 consumers. Due to the interruption of the power supply, the electrically driven pumps and equipment of both the water utilities and the WWTPs ceased working as there was a shortage of emergency power generators with the required capacity. The pressure in the water supply systems fell and varied significantly. The greatest cause for concern in regard to public health considerations was not so much the stagnant water in the piped distribution systems as the possibility of groundwater and wastewater feeding back into the water pipes. In compliance with the provisions of Government Decree 201/2001 (X. 25.) on the quality requirements of drinking-water and the associated control procedures, the municipality and the water supply company notified the population that, on the recommendation of the National Public Health Service (NPHMOS), any water intended to be used for drinking or cooking should be boiled until the time when the water pressure had been fully stabilized. In several settlements water tankers had to be used to ensure sufficient water supply, which proved difficult under the snowy conditions. The measure remained effective until the achievement of a negative bacteriology result in the water samples taken after the stabilization of the water pressure confirmed the restoration of water supply.

Under the difficult ground conditions the restoration operation took more than 72 hours.

Both water supply and wastewater treatment are of key importance for public health, as the lack of them may constitute a considerable health risk, and therefore the provision of an adequate back-up power supply based on a satisfactory number and capacity of emergency generators, to cover for the eventuality of power cuts caused by extreme weather conditions, is an important factor in the vulnerability assessment.

Source: compiled by representatives of the National Institute of Environmental Health, Hungary

Chapter 8

ADAPTATION MEASURES FOR DRAINAGE, SEWERAGE AND WASTEWATER TREATMENT
Ms Doubravka Nedvedova
Ministry of the Environment, Czech Republic

Flooding of urban wastewater treatment plant, Prague, Czech Republic, 2002
© Water Research Institute (Prague)

Adaptation Measures for Drainage, Sewerage and Wastewater Treatment

8.1 Key Messages

Managers and operators of sanitation systems follow a similar approach to those responsible for water supply systems in adapting to extreme weather events. However, they face some specific challenges.

- To assess the risk level of potential impacts of weather events on drainage/sewer systems and WWTPs a punctual analysis should be performed for each element of the systems and under different circumstances of floods and intense rain, drought and prolonged water scarcity, increased temperatures and heat waves.

- Generally speaking, extremely intense rainfall and river flooding are characterized by a primary risk for public safety, while heat waves or extended droughts generally imply a delayed secondary effect on drainage and wastewater systems, specially in urban areas.

- Weather-sensitive design criteria are the primary means for climate proofing in new drainage networks.

- In designing adaptation measures for sanitation systems in extreme weather events it is also necessary to know that every extreme hydrological situation causes fluctuations in pollutant concentrations in wastewater inflow to the WWTPs and thus affects the efficiency of the treatment process. The differences in biochemical load cause problems in different technological sections and related treatment processes.

- In existing networks, the highest hydraulic capacity should be assured by undertaking periodical maintenance and cleaning of the most significant nodes of the network.

- With small networks and limited budgets, decentralized systems face different constraints to climate proofing management and they should have close links with the main environmental authorities and even agreements with centralized systems' managements for emergency interventions.

- The precondition for handling an emergency well is skilled staff able to recognize the danger, analyse the risk and respond properly. Staff should be properly trained and the system regularly tested for emergencies. Another important priority is good communication among everyone involved – system operators, owners, state administration, river basin authorities, the management of official rescue systems and all other stakeholders.

8.2 Climate Change Impacts on Drainage Systems, Sewer Systems and WWTPs

Climate change inevitably leads to changes in weather conditions in affected regions. Increases in average temperature can be expected, and a consequent decrease in both the amount of precipitation and its seasonal distribution. So, for instance, there may be long periods without precipitation followed by extraordinarily strong flash rains.

When speaking about the impact of climate change on the drainage/sewer systems and WWTPs it is important to consider changes in quantity and timing of precipitations (rainfall intensification, flash floods, long dry periods), air temperature, sea level, and a higher frequency of extreme weather (decrease in return period of extreme events).

Storms, heavy rainfall events and a higher frequency of flood events require:

- quality protection of drainage systems, sewer systems and WWTPs against high peaks of hydraulic load;
- additional stormwater storage in retention reservoirs and stormwater tanks;
- treatment of the first flash stormwater containing high concentrations of pollutants.

Prolonged periods without any rainfall lead to:

- a lower wastewater discharge and the consequent accumulation of solid waste sediments and encrustation in sewers that can clog them;
- a decrease in wastewater flow and unpleasant odour from water rotting in the system;
- an increasing population of rodents associated with increased quantities of sediments and solid waste in pipelines;
- a growing risk of disease dissemination;
- saltwater intrusion, especially in coastal agricultural areas, causing the degradation of sewers and affecting the quality of wastewater.

Increasing air temperature affects the processes of wastewater treatment, especially because:

- lower oxygen solubility in water can lower the efficiency of active sludge compartment, which leads to higher consumption of compressed air in biological treatment processes for the same treatment effect;
- higher dust concentrations raise the costs of air filtration;
- faster biological processes (activated sludge process, digestion) are faster because of the higher air temperature, and
- sludge dewatering processes are also more efficient;
- the costs of heating up anaerobic digesting facilities falls.

One of the first steps in improving knowledge of the impacts of weather events on drainage/sewer systems and WWTPs is the assessment of the risk levels related to the different events. In particular, such events as extremely intense rainfall and river flooding are characterized by a primary risk for public safety, while heat waves or extended droughts generally imply a delayed secondary effect on drainage and wastewater systems.

The decreasing dilution capacity of the receiving water bodies is an important aspect of climate change because the longer droughts to be expected would lead to a significant decrease in river flow. Yet the pollution load of wastewater will be higher because users will have saved water during dry periods or because of high pollution concentrations in the first flash stormwater after the dry period. Higher concentrations of pollutants in the wastewater together with the decreased dilution capacity of the recipient bodies would require an increased purification effect from the WWTPs and might even prompt the imposition of stricter limits for discharges in order to keep the current quality level of surface waters. So careful monitoring of water quality in WWTP discharges and in recipient bodies is essential during droughts.

When designing measures for the adaptation of sanitation systems to extreme weather events it is also necessary to remember that every extreme in hydrology causes fluctuations in pollutant concentrations in wastewater inflow to the WWTPs and thus affects the efficiency of the treatment process. The differences in biochemical load cause problems in different technological sections and related treatment processes.

8.3 Adaptation Measures to Urban WWTPs Before and During Droughts

From the point of view of the drainage of urbanized areas, in contrast with extreme rain and river floods (primarily a risk for the safety of the inhabitants), extreme droughts endanger the environment in particular and have a lower immediate impact than other extreme weather events.

However, the necessary management, maintenance and operational measures for the mitigation of the adverse effects of extreme drought on sewer systems and urban WWTPs (UWWTPs) should be mentioned. The following may be regarded as the main consequences of an extremely dry period:

- reduction of the ability of sewage to pass through the pipe because of encrustation and sediments, and lengthening of the time of stay of wastewater in the system caused by partial or full blocking of pipeline;

- deterioration of water organoleptic properties – bad smell from the sewage system, as well as from the first section of the treatment plant;

- increased occurrence of rodents and other vermin in and around the sewage system;

- possibility of infection spread.

Preventative measures must be organized in time, and planned as a part of standard plant operation and maintenance linked to average rain-free periods in the affected areas.

Establishing effective communication channels with city cleaning organizations is especially important during drought periods to prevent inputs from roads and pavements into the drainage and sewerage systems. Communication with other stakeholders is also important to inform them on behaviours that would limit stress to the drainage and sewerage systems.

8.3.1 Maintenance of sewer systems during an extremely long dry period

During a long drought, periodic checking and cleaning of electro-mechanical equipment and of sewage system pipelines and accumulation tanks should be carried out, to ensure their full hydraulic capacity and to prevent the accumulation of solids. In the case of big combined sewer systems, regular and frequent cleaning of roads and pavements (in urban areas), and sewer system inlets and manholes is also important. During long dry periods, it is recommended that the sewer system pipelines should be rinsed with service water (water treated in the biological stage of UWWTP, to avoid wasting water treated to public supply standards). For rinsing urban surfaces, completely safe disinfected water has to be used, according to public health regulations. The pipelines should be water proofed and submitted to regular checks. This is especially important in coastal areas to prevent infiltration of sea water into the sanitation system. In addition to the additional hydrological load, the salt could damage the pipe network.

The availability of simultaneous models used to identify the most critical points when considering sedimentation and hydraulic capacity is increasing. Extensive monitoring and measurements are necessary for calibrating these systems. It is also necessary to remember that mathematical models illustrating the rate of flow through sewage systems are less accurate in establishing minimum flow rates. In some cases sufficient information on possible sediment formation during extremely low rates of flow may be obtained only through monitoring the current status of pipelines. As a general criterion, it is best to verify that the flow speed of the wastewater is >0.5 m/s to ensure the sewers retain a self-cleaning capability.

All usual as well as exceptional maintenance procedures should be defined in the operational maintenance, crisis and emergency plans, and part of those should include a description of communication with stakeholders. For example, the use of domestic grinders for disposal of moist domestic waste should be restricted in such conditions. Appropriate staff training is also necessary so they can identify risk, and respond to it. In the case of extensive city sewage systems, maintenance of roads and sewer and drainage networks is usually carried out by several companies, and so a joint maintenance plan valid for all participating companies must be drawn up.

8.3.2 Operation of UWWTPs during extremely long dry periods – changes in hydraulic and pollution load

UWWTPs designed to ensure a sufficient level of wastewater treatment from combined sewer systems are usually designed to be able to maintain high efficiency even during marked fluctuations of rate of flow and pollution. In periods of longer drought, and in subsequent periods of torrential rain, these fluctuations are even higher. Differences between the level of the UWWTPs' pollution load in workday peaks, or during washing out of sediments from the sewage system, and the level of their pollution load overnight in the rain-free period, are extreme. A treatment plant's technological equipment should be designed taking into account this fluctuation of pollution as well as hydraulic load.

An important part of a UWWTP's technological adaptation for periods of extreme drought should be the possibility of regulating the amount of dissolved oxygen in activation tanks (the amount of air pumped into them), in order to ensure that conditions of operation of the UWWTP's biological section during an extreme pollution load are similar to those during normal operation. Online oxygen-measuring sensors would consequently help the regulation of the process.

8.4 Adaptation Measures Before and During Floods

The knowledge of weather change indicators and provision of information on current weather conditions in the relevant area are fundamental aspects of strategies for protecting sanitation systems against flash rains and floods. This enables an appropriate response by the UWWTP's management to local extreme weather changes, showing increased rains or decreased periods between flash rains, and prevents overloading of sanitary systems.

The potential vulnerability of the system during heavy rains and floods is related to the spatial variability of changes in hydrological conditions, type of drainage and sewer system (combined or separate sewerage system, pipeline location above the ground in more temperate countries), type of technological process used in wastewater treatment (biological, physico-chemical), and also to the structure of the drained area (centralized systems, or decentralized local systems). Adaptation measures should be defined and implemented specifically according to the weak spots of the network.

8.4.1 Centralized drainage/sewer systems and UWWTPs – preventive measures

Centralized systems are often characterized by a large drained area with a high share of built-up area, served by one central main UWWTP with a dominant role and possibly several smaller facilities in the city outskirts. The vulnerability of centralized systems can be seen in the function of the central UWWTP. When an emergency occurs and the central UWWTP is out of order, all the wastewater from the area is discharged into the receiving water body in one spot without any treatment. This could have a significant impact on surface water quality.

Centralized urban sewer networks were often built step by step as the cities grew. Rainwater from built-up areas (roads, pavements, etc.) was drained together with the wastewater from households and industries. This type of combined system worked satisfactorily until a heavy rainfall occurred, when the capacity of the system was insufficient. Because of high concentrations of pollution in the first flushed water, combined sewer overflows (CSOs) can be potential sources of pollution for the receiving surface waters. The problem can be overcome by installing retention tanks for the storage of extra flush water, which can be treated in the UWWTP when the storm is over.

In modern housing developments or in rural areas where no sewerage system existed before, separate drainage systems used to be installed. Separate drainage systems have two pipelines (branches), one for sanitary wastewater, which carries its content to the UWWTP, and one for surface water (rainwater), ending directly in the receiving water body. This type of sewer system is effective in controlling flows to the UWWTP. Problems could arise with misdirected connections and also with high pollution of surface water from the first flush. In many European countries the water volume collected during the first two or three hours of a precipitation event is redirected from the drainage system to primary treatment in order to reduce the content of sediments washed out from the surface of roads and roofs. The primary sedimentation in the highly polluted first flush water can be improved by chemical-physical treatment (coagulation-flocculation)

In adapting to climate change, experienced as increased rains or flash floods, the preparedness and immediate feedback from the managers of sewers and UWWTPs play a key role in preventing the systems flooding. The age of urban facilities can be a significant limitation. The structure of the network may be too poor to allow it to be adapted to changed hydraulic and chemical loads without problems. The adjustment of the whole hydraulic capacity of the system can be economically impossible. In these cases the risk of flooding of the system can be mitigated but not fully eliminated.

The highest possible hydraulic capacity of existing sanitation systems can be achieved mainly by periodic routine maintenance of the system and its important nodes, including precise cleaning of drainage/sewer pipelines and fittings, and other parts of the system to avoid overfilling or congestion. The pipelines must be kept waterproof and impermeable to prevent the infiltration of wastewater to groundwater as well as to prevent the penetration of water from outside to the sewers during flooding. Mechanical and electric parts of pumping stations should be regularly checked and maintained. Apart from ordinary electro-mechanical equipment, a spare flood pumps should also be available for emergencies. Alternative electricity sources should be also prepared because power failures can often occur during floods as well as in other weather extremes.

It is advisable that treatment plants' managements develop and regularly update maintenance, crisis and emergency plans for sanitation systems for both floods and droughts which provide operators with management rules and levels of critical indicators, enabling them to recognize the potential threat and to be able to respond on time and correctly to the potential emergency. Especially in the case of large systems, draining extensive urban areas where different parts of the system are managed by different operators, common maintenance and emergency plans are desirable to secure lasting clarity about the whole system.

The fundamental assumption of a well-prepared crisis and emergency plan is intimate knowledge and detailed mapping of the system. The best way would be to use an easily available detailed plan in geographic information system (GIS) form. The modelling of a hydrological regime of adjacent surface water bodies in emergency situations necessary for effective emergency planning can be performed by the UWTTP operators, but what is more effective is cooperation with other specialized bodies such as forecasting agencies, river basin authorities and so on.

Doing one's own hydraulic modelling of the sanitation system in emergency situations for current and expected weather conditions is highly recommended. It allows UWWTP operators to define weak spots in the system and to propose effective measures in planning procedures. The use of simulation models of stormwater runoff in complex networks has recently increased due to their better accessibility and reliability. The accuracy of simulated scenarios primarily depends on a deep knowledge of the network and precise model calibrations based on measured data. It is necessary not only to develop a mathematical model of a sanitation system, but also to calibrate the model under a certain set of meteorological conditions and link it to the real-time meteorological data collection system. The operators can use the results of this model to prepare the sanitation system and UWWTP for the flow and load fluctuations to minimize the impact on the system and the recipient.

The precondition for handling an emergency well situation is skilled staff able to recognize the danger, analyse the risk and respond properly. The staff of the facility should be appropriately trained and the system regularly tested for emergencies. Another important aspect is good communication among all actors, system operators, owners, state administrations, river basin authorities, managers of official rescue systems, and all other stakeholders.

8.4.2 Decentralized and community-based sanitation systems – preventive measures

Some issues concerning centralized sanitation systems mentioned above can be taken into account also when considering decentralized and community-based systems. But the situation in this case remains different:

- decentralized systems are characterized by small networks and facilities with limited budgets;

- in countries with warm winters the sewage systems of small settlements used to be located above the ground, which makes them more vulnerable to extreme weather;

- the equipment of small networks and facilities is usually not at the highest technological level, so during emergencies it is less effective;

- the use of simulation models and telemetry systems may not be affordable, so experience and knowledge of the network as well as of all its equipment and of its reaction to past critical events becomes more important (periodic surveys, a register of critical situations during extreme events, etc.);

- extreme weather events do not very often concentrate their effects on a smaller area, although when it happens it can hit a major part of the system; response to the impact is therefore less flexible in terms of operational, hydraulic and treatment capacity.

The impact of stormwater on decentralized systems can significantly increase water flow fluctuation during rainfall events, so the adoption of measures mitigating the water inflow to the UWWTP is highly recommended, for example support of safe infiltration of rainwater directly on to the surface, or separation between sewer and rainwater drainage networks. This measure is useful also in the case of massive snow falls when the anti-cryogenic chemicals are being used on the roads. These chemicals can harm purification processes in very small UWWTPs with a low content of water and sludge. The separation of drainage and sewers is possible only if suitable natural receiving bodies for direct water discharges are available. In the case of a combined sewer system the amount of anti-cryogenic chemicals should be controlled by a special regulation.

Where the sewage pipelines are located above the ground it is recommended that they should be secured against damage caused by stormwater where necessary, and against landslides in hilly areas.

Operators of decentralized systems should closely cooperate with the state environmental authorities and with operators of the main centralized sanitation systems to use their experience and equipment (it is advisable to sign agreements on cooperation in emergencies).

For those systems cost–benefit analysis could be employed to decide whether some way of connecting to a neighbouring large centralized system can lead to greater overall efficiency.

Many rural settlements are not equipped with sewers and treatment plants. Wastewater disposal from households is achieved by small home WWTPs, and septic tanks or simple cesspools are used for biological waste. When small home WWTPs are being designed for wastewater disposal it is necessary to take into account the location of remote houses in flood-prone rural areas.

8.4.3 Centralized drainage/sewer systems and UWWTPs – protective measures during floods

Even if weather forecasts are nowadays detailed and precise, the assessment of the extent of a flash storm and its impact on the sanitation system is unpredictable. This is valid especially for extensive combined sewer systems. Flooding does not only mean a temporary increased level of river or other surface waters during which the area in the proximity of the riverbed is flooded. The damage can be caused also by the water not being able to flow naturally away from any area, or by the volume of rainwater exceeding the capacity of the drainage system for a longer time.

In long highly fluctuating rivers, flood waves generated by intense rainfall can propagate quite rapidly, with potentially dangerous effects on sanitation systems along the river as well as downstream in areas not directly affected by storm events. During floods surface water from the river can enter the sewer system through hydraulic connections between the river and the sewer, for example CSOs, weirs, outlets, and leaky sewage conduits. UWWTPs can also be flooded directly from the river or from backwater in the surrounding area.

Reasons leading to the flooding of UWWTPs can be insufficient flood protection, failure in the system or the magnitude of the flood wave. Damage can be caused also by human factors if duties assigned in an emergency plan are neglected. Good management of a sanitation system therefore requires deep knowledge of the hydrological regime of adjacent surface water bodies. The operators should have good communication with river authorities, dam and sluice operators, and flood protection bodies along the river and should be prepared to immediately install components of a flood mitigation and protection system (mobile flood protection walls, pumping devices, etc.).

Floods always affect all the hydraulic capacity of a drainage system. Natural gravitational discharge from the drainage system or from a UWWTP is gradually disabled, and it is necessary to start operating reserve pumping stations. That is why full operability of ordinary as well as emergency electromechanical equipment (flood pumps) should be secured. Electric generators should also be on standby and their connection to the key facilities of the plant (pumping stations, recirculation, mixing and aeration) should be available as break-outs are frequent. The skilled staff should be ready to assume all the responsibilities which the crisis and emergency plan demand. The vulnerability of the system is then strongly influenced by the number of pumping stations available to pump the excessive water back to the receiving water body. UWWTPs located on flat land are more vulnerable than those connected to a gravitation sewer.

The full functioning of UWWTPs and pumping stations should be ensured for as long as possible by continual checking/maintenance of electro-mechanical facilities as well as of the structural components of the system. Fundamental maintenance activities within the system are solid waste disposal and cleaning of sewer conduits, stormwater tanks and other objects to maintain the maximum hydraulic capacity of the system.

For large systems, a central operational unit should be in charge of supervising the response of the whole network, directing special maintenance teams and alerting public and other stakeholders involved in the case of risk to public safety (sewer back-flooding of urban areas, possible contamination of sensitive areas with wastewater, pollution of water bodies and groundwater sources, etc.).

Simulated hydraulic models can be used to define appropriate ways of working during the extreme event in real time. The epicentre of extreme rainfall events is usually confined to a small area, so it could be possible to use the residual hydraulic capacity of drainage/sewer networks by regulating the flow or redistributing stormwater towards branches facing less demand. This kind of measure requires interconnections between different parts of the sewerage network, installation and proper maintenance of automatic control system elements within the system, and a well-calibrated centralized system of real-time control (RTC).

The protection of the UWWTP must be based on the assessment of the real level of risk so that its operation will be maintained and the flooding of vital equipment and pollution of the surface water prevented. There are several pieces of practical advice on how to protect a UWWTP against flooding:

- the building of a flood protection embankment around the UWWTP and sewerage facilities, and the possible construction of mobile flood protection walls;

- closing anti-backflow devices (valves, gates, sluices…) and using pumping stations to protect the systems against the back wave from the recipient water body;

- locating essential UWWTP compartments above the level of the flood with long return period (operation of this solution is more costly, because the whole inlet has to be pumped up even during dry periods);

- using the principle of "containment" – crucial technology and electro-technical equipment is placed in one protective steel container or concrete structure, ensuring the safe operation of the facilities inside during the flood;

- removing and storing dismountable equipment from the UWWTP and the sewer facilities to prevent damage;

- removing all chemicals and other potentially dangerous contaminants from the area of the UWWTP and sewer facilities to prevent additional pollution of surface and groundwater.

During the flood the UWWTP should be kept in operation as long as possible even if most of the wastewater is treated only partially. A high dilution capacity of the receiving water body during flooding makes it possible to keep an acceptable quality of water, even if only part of the incoming wastewater, for which the UWWTP is designed, is treated completely.

If flood protection is overwhelmed and the situation does not allow operation of the UWWTP any more, electrical devices (pumps, compressors, and other electrical devices needed for the restoration of the treatment plant) should be removed first to avoid any damage caused by rising water.

8.4.4 Decentralized and community-based sanitation systems – protective measures during floods

Very often the operators of decentralized sanitation systems are not well enough prepared to handle emergencies without the help of professionals, mostly because of a lack of reserve equipment (pumps, tanker trucks, etc.) and sometimes also because of insufficient expertise among the staff. Good cooperation during the emergency with the state administration and with operators of central UWWTPs or other organizations owning the necessary equipment can help to overcome this problem.

Activities that have to be undertaken during and after the flood (until total restoration of the system) are:

- to avoid the release of contaminated water and especially contaminated sludge from cesspools and septic tanks into surface and groundwater;

CASE STUDY 13 : SCENARIOS OF WATER ENTERING UWWTP AREAS DURING FLOOD, CZECH REPUBLIC, 2002

These are some examples of problems faced by UWWTPs in the Czech Republic in 2002.

During the big 2002 flood in the Czech Republic several tens of large UWWTPs were flooded. The scenarios were different and they can be divided into four categories:

- overflowing of flood protection embankments because of rising water level in the river (central UWWTP in Prague);
- inundation of WWTP area by surrounding floodwaters because of insufficient protection upstream (UWWTP in České Budějovice);
- flooding of WWTP because of over-large inlet of water from the river which entered drainage and sewer systems through insufficient anti-backflow devices (UWWTP in Teplice, UWWTP in Ústí nad Labem);
- flooding from backflow water coming from downstream.

Source: compiled by Vaclav Stastny, Water Research Institute Prague, Czech Repub.

- to protect drinking-water resources, wells and boreholes from contamination by polluted flood water;

- to keep the sanitation system working as long as possible;

- if necessary to remove all electrical devices from pumping stations and small domestic WWTPs endangered by flooding so that they can be used for a smooth re-start of the system after the flood;

- after the flood to remove all sludge from cesspools and septic tanks, to have it treated in the nearest UWWTP and to start to operate it as soon as possible;

- to rinse flooded sewers with high-pressure water if needed and reset the operation of the local sewer system and UWWTP as soon as possible.

In rural areas it is very important to observe basic hygiene requirements: to prevent a mass escape of dangerous bacteria to surface and groundwater, and to keep the sanitary system operating as long as feasible, or in a state that allows it to be restored as rapidly as possible to an operational condition.

8.5 Restoration of the Sewerage System and UWWTP

Once the extreme weather event is over, damage assessment is the fundamental part of the restoration process. It must be carried out as quickly and thoroughly as possible. It is necessary to identify the most affected areas, and to define priorities and the help that is needed, in order that management of the restoration is logical and methodical, not chaotic. Further, it is necessary to carry out a first analysis of what actually happened during the flood, and to roughly assess its impacts and direct causes (i.e. in particular to identify inadequate flood protection measures in order to improve them in the future). Later on it will be necessary to draw up a more comprehensive estimate of risks and an analysis of the individual aspects of the event and its effects. For this secondary process, it is important, in particular:

- to specify all critical elements in the system, in relation to the flood;
- to develop an expert analysis on the basis of newly collected data;
- to design new measures;
- to improve infrastructure for any possible future extreme event.

8.5.1 Regaining and restart of drainage/sewer network operation

In particular the following activities need underlining as basic operations ensuring the restoration of functions of the drainage/sewer system after the extreme event:

- restart normal operation – through a special service team, working uninterruptedly 24 hours a day, ensuring free passage through conduits and checking and restoring the main pumping station collectors in the sewer network;

- perform uninterrupted centralized monitoring of the functions of the drainage and sewer system main junctions, accumulation tanks, pumping stations, etc.

The long-term impacts of possible service failures on the local inhabitants should be mitigated. In this critical period, it is necessary not to neglect relations with the public and stakeholders, which can be encouraged through prompt provision of information and dealing quickly with all complaints.

8.5.2 Regaining and restart of the UWWTP operation

Just as with the restoration of sewer network function, the restoration of UWWTP operation after a flood has to be started immediately after the danger has passed.
The procedure for the start of UWWTP operation is as follows.

- General primary measures include:
    - making all areas of the UWWTP accessible and documenting damage;
    - gradual removal of dirt, disinfection and cleaning of buildings and technological equipment (if necessary);
    - drawing up a plan for the restart procedure;
    - assessing the stability of buildings and structural compartments.

- Technological measures include:
    - ensuring the electrical energy supply;
    - making wastewater distribution systems and the bypass systems of the UWWTP passable.

- Technological start of the UWWTP operation includes:
    - start of pre-treatment (cleaning grating, grit chamber, grease trap) operation, for the present bypassing the sedimentation and biological stages;
    - start of sludge management operation (at least storage reservoirs), the necessary precondition for the start of primary sedimentation (due to the usual high content of solids in the treated water, it is necessary to protect the existing pumps by drawing away most of the sediments);
    - start of operation of mechanical treatment stage, and chemical precipitation in sedimentation tanks (the preconditions for this are functional pre-treatment and sufficient capacity for sludge storage);
    - removal of solids from activation basins (if this is not possible, mobile mixing in activation basins is necessary);
    - start of operation of the aeration system, and subsequently gradual putting into operation of the biological stage (its precondition is functional primary sedimentation);
    - gradual putting into operation of automated management systems, preceded by manual dispatcher control.

The most difficult part of the procedure is gradually putting the biological treatment stage into operation. During the flood, activated sludge is usually washed away from the tanks, or its decay takes place (a similar situation occurs by anaerobic sludge stabilization in sludge digestion tanks). It can be done by:

- inoculation with sludge from another UWWTP;

- "cultivating" new active sludge without inoculation;

- using the original sludge(although decayed) which has not been washed away, thanks to the sedimentation in the biological stage tanks after aeration switch-off and before the inflow of flood water (the best method).

According to experience from the Czech Republic in 2002, reaching performance of the biological stage which is comparable with the original standard takes several months, regardless of the method used for putting the activation tanks into operation. In southern countries, where warm weather prevails, the process could be shorter.

To speed up the process of returning a UWWTP to work after a flood, it should be modified according to the level to which the UWWTP was affected, and also according to how it was shut down before the flood. For example, if the electrical equipment was protected by means of dismantling, cleaning it will not be part of the process, but re-assembling it will be. If some of the UWWTP's stages were not affected thanks to the success of flood protection measures, the procedure of putting it back into operation is shorter and less complicated. However, the sequence of putting the individual stages into operation always has to be kept.

8.6 Specific Issues of Industrial WWTPs

The functioning of industrial WWTPs (IWWTPs) can be especially threatened during floods. Flooded and washed-out IWWTPs can represent enormous danger for people and for the environment because of the possible presence of hazardous substances (Case study 14).

As industrial facilities use different technologies for wastewater treatment according to the quality of treated wastewater and the techniques chosen, the flood protection strategies and damage recovery in IWWTPs will require specific approaches. Besides the basic principles applied by UWWTPs, the protection of IWWTPs against floods will require specific crisis and emergency plans, including accident emergency warning systems connected to the national one, and individual detailed analyses of technology/technique, equipment, and chemicals used and discharged. Industrial wastewater, sludge and chemicals stored and used in the process have often to be treated as hazardous and so regulations on hazardous substances have to be scrupulously respected. Special training for staff is essential.

Flood protection measures in IWWTPs and restoration activities should take into account special technologies used as a part of industrial treatment processes, which require special approaches not normally needed in UWWTPs. Beside the protection of special electrical equipment (e.g. for electro-flotation), safe intermediate storage of contaminated wastewater, sludge and other chemicals should be designed (for example safe tanks, flood-protected industrial lagoons for liquid hazardous wastes, etc.).

During a flood an emergency can develop not only as a threat to surface or groundwater, but also to other parts of the environment (through contamination of soil, toxic gas emissions, etc.), so operators should be trained and prepared to respond to these situations as well.

During the flood it is advisable to:

- ensure the good operation of pumping stations, in order to make sure they keep working so as to avoid releases of untreated industrial wastewater into surface or groundwater;

- protect each compartment of an IWWTP against flooding, including temporary toxic sludge storage;

- install (temporarily or permanently) mobile units for wastewater treatment – these can operate above the level of flood water to avoid untreated water being released into surface or groundwater;

- keep the internal warning system which monitors defects in the IWWTP in operation even during the flood;

- be prepared to clean all contaminated equipment and dispose of hazardous toxic wastewater and sludge. For this purpose a recovery plan drawing on experience to date can be developed. Wastewater or sludge with hazardous substances should never be released into surface or groundwater.

The guidelines for restoring IWWTPs after flood are basically the same as for large UWWTPs – the accessibility of the area, a damage assessment, the re-establishment of electrical power, cleaning sewer conduits and step-by-step restoration of all treatment processes (from preliminary treatment to automatic control systems). Obviously, if some of the processes were operating during the flood, the restoration process will be much faster.

8.7 Summary

Climate change brings about extreme weather events, predominantly in the form of extreme hydrological phenomena – for example, long periods without precipitation, accompanied by higher temperatures and followed by extraordinarily strong flash rains.

When discussing the impact of climate change on drainage/sewerage systems and WWTPs it is important to consider changes in quantity and timing of precipitations. Prolonged periods without any rainfall lead to a decrease in wastewater

> ### Case study 14 : Damage Caused by the Flooding of the IWWTP in Rozoky, Czech Republic, 2002
>
> This IWWTP has an anaerobic mechanical-biological treatment process. The purification is completed in a UWWTP. The WWTP can treat 72 000 P.E. and the inlet is 840 m³/day. During the 2002 extreme flood it was totally inundated. The water level in the receiving river overtopped a 6 m-high flood protection embankment with a 1.4 m-thick overflowing jet. The floodwater pressure even lifted the gas tank in the basement, the building housing the engine room was damaged, the tank for chemical injection floated 100 m away and other tanks remained in place tethered only by the pipeline fittings. All electro-motors and control systems were destroyed. The total restoration costs reached 29% of the budget for the previous reconstruction in 2000. Yet the WWTP was operating again after just three months.
>
> After the experience gained from the 1997 and 2002 floods in the Czech Republic, flood protection measures on IWWTPs should be an important part of crisis and emergency planning, which is dealt with when the administrative permission for IWWTP is being issued according to the Czech regulations.
>
> *Source: compiled by D Nedvedova, Ministry of the Environment, Czech Republic.*

discharge, the pollution load of which could be higher due to the application of saving strategies in water supply and water use. This is accompanied by accumulation of solid waste sediments and incrustation in sewage conduits, causing conduit clogging, an unpleasant odour from water rotting in the system, an increase in the rodent population and a growing risk of spread of diseases. Hence, permeability and cleanliness of the system should be regularly checked and maintained. In coastal areas saltwater intrusion should be prevented as it causes degradation of sewer conduits and affects the quality of wastewater.

Rising air temperature affects wastewater treatment processes. Biological processes and sludge dewatering are more efficient with identical hydraulic and pollution loads, but lower oxygen solubility in water leads to higher consumption of compressed air in the biological treatment process for the same treating effect, and higher dust concentration causes higher air filtration costs. Operators should remember that extreme hydrological conditions bring about fluctuations in pollutant concentrations in wastewater inflow to the WWTP, which can cause problems in the treatment process.

One of the most significant impacts of rising temperature and draught is decreasing dilution capacity of the receiving water bodies. Expected longer dry periods would lead to significant decreases in river flow. This would require the application of more stringent emission limits for the water discharged from the WWTPs, accompanied by careful monitoring of its quality.

Long draught periods have a lesser impact on sanitation systems than intense rainfalls and flash floods. It should be borne in mind that a long dry period can be followed by flash rains and then the combination of both natural phenomena influence the sanitation system in synergy. Extreme long dry periods can cause sewerage incrustation. The following flash rain then washes out pollution from the sewerage system and overloads the WWTP hydraulically, as well as in terms of pollution levels. This is why water should be treated during flash rains in the WWTPs and, if possible, not released directly to the recipients. To this end, it is necessary to equip WWTPs with a special reservoir to retain this increased level of pollution in wastewater, in order to have it treated later, while at the same time reducing the negative impact of flash rains on the system.

In the case of storms, heavy rainfall and higher frequency of flood events, drainage/sewerage systems and WWTPs must be protected against flooding and high peaks in hydraulic and pollution loads. Regular maintenance of the system and its important nodes – including careful cleaning of sewer conduits and fittings – can ensure the highest possible hydraulic capacity of the sanitation systems. During the event the system should be kept in operation as long as possible. It is advisable to have available spare pumps and emergency electrical generators. Chemicals and other dangerous substances must be stored in a safe, dry place. After the event the system should be put into operation as soon as possible, on a step-by-step basis, dependent on the level of damage. Communication with the public should be maintained at all times.

Good knowledge of weather change indicators, modelling of the hydrological regime of nearby water bodies, use of simultaneous models to identify the sanitation system's most critical points and smooth information flow in the affected region before and during an extreme event all serve to facilitate an appropriate response on the part of the WWTP's management team to the local climate change (to be proven during periods of increased rainfall or by decreased periods between flash rains), and prevent overloading and engorgement of the sanitary system.

It is advisable to develop and update maintenance and emergency plans for sanitation systems. In the case of large systems in particular – including drainage of roads and urban areas in which different parts of the systems are being operated by different actors – common maintenance and emergency plans are desirable, and could comprise a strategy for communication with stakeholders and public. It is advisable to have a well-trained staff team recognizing and responding to emergency situations.

8.8   Checklist

See Table 27 (For a checklist for adaptation measures for drainage and sewerage systems).

Case Study 15 : Sewerage Network and Sanitation Planning, Management and Recovery in Case of Extreme Events, Belgium

Planning of wastewater infrastructure for extreme events

In the Flemish Region, the use of hydrodynamic models to simulate the effects of rain events on the sewer networks has for many years been widely integrated and applied in the processes of planning, designing and construction of new sewer networks, as well as in making best use of the existing sewer networks. Extensive drought periods do not occur in this Region in a way that has a critical effect on the drainage and sewerage systems. However they can have an environmental impact when followed by intense rainfall events, since in the Region the majority of the sewer networks consist of combined sewerage systems.

Therefore the collecting systems are constructed following these standards:
- allowing extensive volumes (at least six times the volume of the dry weather flow) of a mixture of wastewater and rainwater to be transported towards the centralised treatment facilities;
- limiting the pollution of receiving waters from stormwater overflows to a maximum of seven days of overflow/year.

In addition, the construction of sewer networks has over the last decade and more evolved into separated systems. This aims at preventing the rainwater from flowing into the sewer networks and at keeping the rainwater where it falls in order for it to infiltrate back into the ground. This enlarges the existing hydraulic capacity of the sewer systems for events of extreme rainfall.

Management of wastewater infrastructure for extreme events

As for management, sewer systems need to be big enough to prevent untreated wastewater from overflowing into the rivers, and treatment plants need to treat sufficiently high volumes of a mixture of wastewater and rainwater when there is intense rainfall. For several decades it has been standard practice in the Region to secure biological treatment of a volume equal to three times the dry weather flow and to provide a primary treatment on an equal supplementary volume. About 10 years ago, a study of Aquafin, the public limited company responsible for the construction and management of the treatment plants and large collecting systems of the Region, showed that the entire volume collected towards the treatment plants can be treated biologically (in secondary and in most cases tertiary treatment, to remove nitrogen and phosphorus) without substantial costs. All renovated and new treatment plants and/or those where this could be implemented without excessive cost now have these treatment facilities. Aquafin also continuously monitors the functionality of the systems of all treatment plants and crucial pumping stations of the sewerage network to prevent ecological damage. This continuous monitoring, with a linked alarm system, enables a quick response in case of extreme events. Since 2008 the Region has been implementing a supplementary control system for the good management of the treatment facilities by monitoring different indicators, such as the continuous functionality of the treatment plants and the pumping stations and their adequate reactions in case of dysfunction.

Restarting WWTPs after flooding (restoration)

In the Flemish Region, the functionality of WWTPs is rarely affected by flooding events. The recovery activities were limited to the restarting of the installations due to power cut-offs.

Monitoring of wastewater infrastructure: storm overflows

The Flemish Environment Agency (VMM) is responsible for the ecological supervision of the wastewater treatment infrastructure. Since 2002, VMM has elaborated a network of measuring stations to monitor the effects of storm overflows on the surface water quality. This storm overflow network consists of 250 measuring stations running on solar energy which monitor overflow infrastructures 24/7 with level sensors and quality sensors (for turbidity, conductivity and temperature).

After a few years of measuring expertise, the following conclusions can be drawn:
- the expected maximum duration of overflows working, namely 2% on a yearly base, was an underestimate; an average of 3.4% has been measured;
- the data from the measuring stations have revealed several bottlenecks, resulting in local adaptations to the infrastructure and further optimization of investments in the sewer system;
- the monitoring network is used as an instrument to support grants or refusals for companies to connect to the public sewer system.

Since 2008 VMM has redirected the measuring stations to critical points like certain treatment plants and pump stations. These will also be monitored by the WWTPs and local/municipal authorities by means of additional flow measurements in order to improve wastewater collection, transportation and final treatment.

Source: M Van Peteghem, Flemish Environment Agency, Belgium

Table 27. Checklist for adaptation measures for drainage and sewerage systems

General preventative and maintenance measures for urban sanitation systems to be prepared for extreme weather events (drought, flood, storm)	Adaptation measures for urban sanitation systems in the case of drought	Preventative, protective and restoration measures for urban sanitation systems in the case of storms and heavy rains (floods)	Preventive, protective and recovery measures for IWWTPs in the case of storms and heavy rains (floods)
• Have a reliable forecast on meteorological and hydrological conditions as well as information on current weather conditions • Detailed mapping of the system should be available, preferably in the GIS form • Use of simulated hydrological models of water runoff based on precise measurements and calibration increase knowledge of vulnerability of the system related to changes in the surrounding hydrological conditions	Extreme droughts do not have an immediate impact on the sanitation system. Specific measures for drought periods are recommended to prevent clogging of the system and allowing the treatment of highly polluted wastewater as well.	Floods represent primarily an immediate risk for inhabitants and property. Besides measures included in part 1, specific measures for protection against flood and mitigation of their impact are recommended	Basic principles applied to UWWTPs are valid also for IWWTPs. It should be borne in mind that an emergency in an industrial plant can affect not only water and its environment, but also other components of the environment.

	Measures to avoid input of solid waste to pipelines to prevent their clogging	Preventative measures against floods	Specific issues of IWWTPs
• Use of mathematical models to illustrate the hydraulic characteristics of the sanitation system helps to identify its most critical points; • For large systems the central operational unit can be in charge to steer and supervise the response of the network • Maintain the system and its important nodes periodically (clean and wash pipelines and tanks to prevent aggregation of sediments, perform regular maintenance of machines, pumping stations and their electric parts) • Develop and regularly update maintenance, crisis and emergency plans based on cooperation of all actors (facilities owners, operators, municipal authorities, road management, river basin authorities, flood authorities, flood forecasting authorities, stakeholders, etc.) • Train the staff for the emergency (drought, flood, storm, wind, etc.) • Involve and inform public • Test the emergency system regularly	• Regularly rinse adjacent pavements and roads (with hygienic safe water) • If necessary ban the use of domestic grinders on the moist fraction of waste • Carry out careful measurement and monitoring for calibration of the models, which are less accurate in areas of minimum rate of flow	• Construction of separate drainage for rainwater where possible • Construction of protective measures against floods (permanent yet mobile walls) • Installation of retention tanks for superfluous flash water (especially in the case of combined sewers) • Having spare flood pumps available for emergencies • Having alternative electricity sources prepared as a power failure may occur during flood or storm	• Flood protection measures must reflect the fact that each industrial plant uses different techniques and technologies as well as raw materials (e.g. chemicals) • Each industrial facility should have its own crisis and emergency plan, including an internal emergency warning system connected to the public one, and a detailed analysis of techniques, technologies, equipment and chemicals used, as well as basic principles of IWWTP recovery • Regulations valid for hazardous substances have to be respected – safe intermediate storage of contaminated water, sludge and chemicals should be used, permitting their consequent safe disposal • Special training of staff covering all possibilities is essential

		Efficient operation of UWWTP	Protective and operational measures for urban sanitation systems in the case of floods
		• The UWWTP must be constructed to be able to maintain high efficiency in the case of increased pollution loads	• The system should be kept in operation as long as possible • Good communication among UWWTP operators, river basin authorities, dam operators, flood protection bodies, hydrological and hydrometeorological monitoring and forecasting institutions should be standard

Table 27. Continued

General preventative and maintenance measures for urban sanitation systems to be prepared for extreme weather events (drought, flood, storm)	Adaptation measures for urban sanitation systems in the case of drought	Preventative, protective and restoration measures for urban sanitation systems in the case of storms and heavy rains (floods)	Preventive, protective and recovery measures for IWWTPs in the case of storms and heavy rains (floods)
	Efficient operation of UWWTP • Adapt the amount of dissolved oxygen in the activation tanks to higher demand during extreme pollution load (if possible by means of automatic regulation) • The UWWTP must be constructed to be able to maintain high efficiency in the case of increased pollution loads • Adapt the amount of dissolved oxygen in the activation tanks to higher demand during extreme pollution load (if possible by means of automatic regulation)	Protective and operational measures for urban sanitation systems in the case of floods **Technical measures:** • on the basis of warning from the forecasting institution or flood protection body, immediately install flood mitigation and protective technical components (flood protection walls, pumping devices, etc.); • location of the most essential UWWTP compartments above the flood level or in tanks is an advantage; • close anti-backflow devices and use pumps to protect system against back wave from recipient water; • all the time, maintain the maximum hydraulic capacity of the system, and prevent solid sedimentation; • bring electric generators to the standby position and use them when necessary; • if needed and possible, remove all endangered dismountable equipment to prevent its damage; • store all chemicals and other contaminants in a safe place. **Protective measures in rural areas during floods:** • in the case of abundant snowfall restrict the use of anti-cryogenic chemicals, which can damage the treatment process in small UWWTPs; • avoid release of contaminated sludge from septic tanks and cesspools to the water; • protect the sources of drinking-water (e.g. wells) from contamination; • after the flood, remove all sludge from cesspools and septic tanks to transport it to the nearest UWWTP.	

Adaptation Measures for Sanitation

Table 27. Continued

General preventative and maintenance measures for urban sanitation measures to be prepared for extreme weather events (drought, flood, storm)	Adaptation measures for urban sanitation systems in the case of drought	Preventative, protective and restoration measures for urban sanitation systems in the case of storms and heavy rains (floods)	Preventive, protective and recovery measures for UWWTPs in the case of storms and heavy rains (floods)
		Protective and operational measures for urban sanitation systems in the case of floods Restoration of sanitation system after the flood: - restart the system as soon as possible; - assess the stability of buildings and structural compartments; - clean and disinfect the affected UWWTP area, including buildings and technological equipment; - verify, restore and monitor cleanness of conduits and pumping stations; - draw up a plan of recovery on the basis of thorough damage analysis. Technical measures: - ensure electric supply; - clear wastewater distribution and bypass systems; - start pre-treatment (bar screen, grease trap), with bypass of sedimentation and biological treatment; - start sludge management operation (at least sludge storage reservoirs), which is a precondition for the start of primary sedimentation; - start the operation of mechanical treatment and possible chemical precipitation in primary sedimentation tanks (a precondition is the functioning of pre-treatment and sufficient capacity for sludge storage); - start the operation of the aeration system and gradually put the biological stage into operation (a precondition is the functioning of primary sedimentation); - gradually put the automatic management system into operation.	

References

ADPC (2000). *Post-disaster damage assessment and need analysis, 2nd draft*. Pathumthani, Asian Disaster Preparedness Centre (www.reliefweb.int/rw/lib.nsf/db900SID/LGEL-5J2N9Z/$FILE/adpc-needs-aug00.pdf?OpenElement, accessed 28 April 2010).

AIDMI (2005). *Community damage assessment and demand analysis*. Gujarat, All-India Disaster Mitigation Institute (Experience Learning Series 33).

Ahern MJ et al. (2005). Global health impacts of floods: epidemiological evidence. *Epidemiologic Reviews*, 27:36–45.

Auger CMS, Lally JM (2008). Acanthamoeba: a review of its potential to cause keratitis, current lens case solution disinfection standards and methodologies, and strategies to reduce patient risk. *Eye Contact Lens: Science and Clinical Practice*, 34(5):247–253 (http://journals.lww.com/claojournal/Abstract/2008/09000/Acanthamoeba__A_Review_of_Its_Potential_to_Cause.1.aspx, accessed 20 September 2010).

Barredo J et al. (2009). No upward trend in normalized windstorms in Europe 1970–2008. *National Hazards and Earth System Sciences*, 10:97–104.

Bartram J et al., eds. (2007). *Legionella and the prevention of legionellosis*. Geneva, World Health Organization.

Bates BC et al., eds. (2008). *Climate change and water. Technical paper of the Intergovernmental Panel on Climate Change*. Geneva, IPCC Secretariat.

Behets J et al. (2007). Survey for the presence of specific free-living amoebae in cooling waters from Belgian power plants. *Parasitology Research*, 100(6):1249–1256.

BIPE (2006). *Analysis of drinking water and wastewater services in eight European capitals: the sustainable development perspective*. Paris, BIPE.

Blair B et al. (2008). Naegleria fowleri in well water. *Emerging Infectious Diseases*, 14(9):1499–1501.

Buitenkamp M, Stintzing AR (2008). *Europe's sanitation problem. 20 million Europeans need access to safe and affordable sanitation. Report of the World Water Week Seminar, Stockholm, 19 August*. Utrecht, Women in Europe for a Common Future (http://www.waterlink-international.com/download/whitepaper_uploadfile_7.pdf, accessed 5 April 2010).

Camberato J (2001). *Irrigation water quality. Update from the 2001 Carolinas GCSA Annual meeting*. Clemson, SC, Clemson University Turfgrass Program (http://www.scnla.com/Irrigation_Water_Quality.pdf, accessed 14 September 2010).

Campbell-Lendrum D et al. (2003). How much disease could climate change cause? In: McMichael A et al., eds. *Climate change and human health: risks and responses*. Geneva, World Health Organization/World Meteorological Organization/United Nations Environment Programme:133–159.

Chase JM, Knight TM (2003). Drought-induced mosquito outbreaks in wetlands. *Ecology Letters*, 6:1017–1024.

Confalonieri U et al. (2007). Human health. In: Parry ML et al., eds. *Climate change 2007: impacts, adaptation and vulnerability. Contribution of Working Group II to the Fourth Assessment Report of the Intergovernmental Panel on Climate Change*. Cambridge, Cambridge University Press:391–431.

Council of the European Communities (1991). Council Directive 91/271/EEC of 21 May 1991 concerning urban waste-water treatment. *Official Journal of the European Communities* (L327/1, dated 22 December 2000) (http://eur-lex.europa.eu/LexUriServ/LexUriServ.do?uri=CELEX:31991L0271:EN:NOT, accessed 5 July 2010).

Council of the European Union (1998). Council Directive 98/83/EC of 3 November 1998 on the quality of water intended for human consumption. *Official Journal of the European Union* (L288/27, dated 6 November 2007) (http://eur-lex.europa.eu/LexUriServ/LexUriServ.do?uri=CELEX:31998L0083:EN:NOT, accessed 5 July 2010).

Council of the European Communities (2000). Directive 2000/60/EC of the European Parliament and of the Council of 23 October 2000 establishing a framework for Community action in the field of water policy. *Official Journal of the European Communities* (L327/1, dated 22 December 2000) (http://eur-lex.europa.eu/LexUriServ/LexUriServ.do?uri=OJ:L:2000:327:0001:0072:EN:PDF, accessed 6 July 2010).

Council of the European Union (2006). Directive 2006/118/EC

of the European Parliament and of the Council of 12 December 2006 on the protection of groundwater against pollution and deterioration. *Official Journal of the European Union* (L372/19, dated 27 December 2006 (http://eur-lex.europa.eu/LexUriServ/LexUriServ.do?uri=OJ:L:2006:372:0019:0031:EN:PDF, accessed 5 April 2010).

Council of the European Union (2007). Directive 2007/60/EC of the European Parliament and of the Council of 23 October 2007 on the assessment and management of flood risks. *Official Journal of the European Union* (L288/27, dated 6 November 2007) (http://eur-lex.europa.eu/LexUriServ/LexUriServ.do?uri=OJ:L:2007:288:0027:0034:EN:PDF, accessed 6 July 2010).

Croci L et al. (2001). Detection of vibrionaceae in mussels and in their seawater growing area. *Microbiologie-Aliments-Nutrition*, 14:161–165.

Danskin W, Crawford S (2008). Managing seawater intrusion using multiple-depth monitoring wells. In: *Proceedings of the 20th Salt Water Intrusion Meeting Program and proceedings book*. Naples, FL, 23–27 June: 49 (http://www.conference.ifas.ufl.edu/swim/papers.pdf, accessed 14 September 2010).

Del Ninno C, Lundberg M (2005). Treading water: the long-term impact of the 1998 flood on nutrition in Bangladesh. *Economics and Human Biology*, 3:67–96.

DePaola A et al. (1990). Incidence of vibrio parahaemolyticus in US coastal waters and oysters. *Applied and Environmental Microbiology*, 56(8):2299–2302.

DePaola A et al. (2003). Seasonal abundance of total and pathogenic vibrio parahaemolyticus in Alabama oysters. *Applied and Environmental Microbiology*, 69(3):1521–1526.

De Sousa OV et al. (2004). Detection of vibrio parahaemolyticus and vibrio cholerae in oyster, crassostrea rhizophorae, collected from a natural nursery in the Coco river estuary, Fortaleza, Ceara, Brazil. *Revista do Instituto de Medicina Tropical de São Paulo*, 46(2):59–62.

Dias E, Pereira P, Franca S (2002). Production of paralytic shellfish toxins by aphanizomenon sp. LMECYA31 (cyanobacteria). *Toxicon*, 38:705–712.

DiSipio E, Galgaro A, Zuppi GM (2007). Contaminazione salina nei sistemi acquiferi dell'entroterra meridionale della leguna di Venezia [Saline contamination in the water systems of the Southern inland lagoon of Venice]. *Giornale di Geologia Applicata*, 5:5–12.

Dragoni W, Sukhija BS (2008). Climate change and groundwater: a short review. In: Dragoni W, Sukhija BS, eds. *Climate change and groundwater*. London, Geological Society:1–12 (Special Publications 288).

Ebi KL (2008). Adaptation costs for climate change-related cases of diarrhoeal disease, malnutrition, and malaria in 2030. *Globalization and Health*, 4:9.

Edet AE, Okereke CS (2001). Monitoring seawater intrusion in the tertiary-quaternary aquifer system, Coastal Akwa Ibom area, Southeastern Nigeria-Baseline data. In: *Proceedings of Monitoring, Modelling, and Management Conference*. Essaouira, 23–25 April.

Edwards et al. (2006). Regional climate change and harmful algal blooms in the northeast Atlantic. *Limnology and Oceanography*, 51(2):820–829.

EEA (2005a). *Effectiveness of urban wastewater treatment policies in selected countries: an EEA pilot study*. Copenhagen, European Environment Agency (Report No. 2/2005) (http://www.eea.europa.eu/publications/eea_report_2005_2, accessed 16 September 2010).

EEA (2005b). *River catchments affected by flooding (1998–2005)*. Copenhagen, European Environment Agency (http://www.eea.europa.eu/data-and-maps/figures/river-catchments-affected-by-flooding-1998-2005, accessed 5 July 2010).

EEA (2007). *Climate change and water adaptation issues*. Copenhagen, European Environment Agency (Technical report No. 2) (http://www.eea.europa.eu/publications/technical_report_2007_2, accessed 7 July 2010).

EEA (2008). *Impacts of Europe's changing climate – 2008 indicator-based assessment*. Copenhagen, European Environment Agency (Report No. 4/2008) (http://www.eea.europa.eu/publications/eea_report_2008_4, accessed 7 July 2010).

Eiler E et al. (2007). Growth response of vibrio cholerae and other vibrio spp. to cyanobacterial dissolved organic matter and temperature in brackish water. *FEMS Microbiology Ecology*, 60:411–418.

EM-DAT (2009). EM-DAT International disaster database [online database]. Brussels, Université Catholique de Louvain Centre for Research on the Epidemiology of Disasters (CRED) (http://www.emdat.be/database, accessed 5 April 2010).

Epstein PR (1993). Algal blooms in the spread and persistence of cholera. *Bio Systems*, 31:209–221.

EUREAU (2008). *Climate change and water and wastewater services. EUREAU Position Paper*. Brussels, European Federation of National Associations of Water and Wastewater Services (http://www.eureau.org/page.php?id=6, accessed 5 November 2008).

Euripidou E, Murray V (2004). Public health impacts of floods and chemical contamination. *Journal of Public Health*, 26:376–383.

European Commission DG Health and Consumer Affairs (2007). *Rapid alert system for food and feed (RASFF) annual report 2007*. Luxembourg, Office for Official Publications of the European Commission.

FAO (2008). *Climate change: implications for food safety*. Rome, Food and Agriculture Organization of the United Nations (http://www.fao.org/ag/agn/agns/files/HLC1_Climate_

Change_and_Food_Safety.pdf, accessed 19 July 2010).

FAO, WHO (2005). *Risk assessment of vibrio vulnificus in raw oysters*. Interpretative summary and technical report. Rome and Geneva, Food and Agriculture Organization of the United Nations and World Health Organization (Microbiological Risk Assessment Series No. 8) (http://www.who.int/foodsafety/publications/micro/mra8.pdf, accessed 5 July 2010).

Fields BS, Benson RF, Besser RE (2002). *Legionella* and legionnaires' disease: 25 years of investigation. *Clinical Microbiology Reviews*, 15(3):506–526.

Fliermans CB et al. (1981). Ecological distribution of legionella pneumophila. *Applied and Environmental Microbiology*, 41:9–16.

Frangano F et al. (2001). *Strategic paper no. 1: case studies on decentralization of water supply and sanitation services in Latin America*. Washington, DC, Environmental Health Project (http://www.phishare.org/files/890_whole%20document.pdf, accessed 5 July 2010).

Fristachi A, Hall S (2008). Occurrence of cyanobacterial harmful algal blooms workgroup report. *Advances in Experimental Medicine and Biology*, 619:45–103.

Funari E, Testai E (2008). Human health risk assessment related to cyanotoxins exposure. *Critical Reviews in Toxicology*, 38:97–126.

Gatt K (2009). Climate migration. In: Micallef A, Sammut CV, eds. *The second National Communication of Malta to the United Nations Framework Convention on Climate Change*. Floriana, Government of Malta Ministry for Resources and Rural Affairs:Chapter 12.

Githeko AK et al. (2000). Climate change and vector-borne diseases: a regional analysis. *Bulletin of the World Health Organization*, 78(9):1136–1147.

Greer A, Ng V, Fisman D (2008). Climate change and infectious diseases in North America: the road ahead. *Canadian Medical Association Journal*, 178(6):Doi10.

Herath S (2001). *Geographical information systems in disaster reduction*. Kobe, Asian Disaster Reduction Centre (http://www.adrc.asia/publications/Venten/HP/Paper(Herath).htm, accessed 6 July 2010).

Hiscock K, Tanaka Y (2006). The potential impacts of climate change on groundwater resources: from the high plains of the U.S. to the flatlands of the U.K. In: *Proceedings of the National Hydrology Seminar "Water Resources in Ireland and Climate Change"*. Tullamore, 14 November:19–26.

IPCC (2007). Summary for policy-makers. In: Parry ML et al., eds. *Climate change 2007: impacts, adaptation and vulnerability. Contribution of Working Group II to the Fourth Assessment Report of the Intergovernmental Panel on Climate Change*. Cambridge, Cambridge University Press:7–22.

Jamerson M et al. (2008). Survey for the presence of naegleria fowleri amoebae in lake water used to cool reactors at a nuclear power generating plant. *Parasitology Research*, 104(5):969–978.

Johns DG et al. (2003). Increased blooms of a dinoflagellate in the NW Atlantic. *Marine Ecology Progress Series*, 263:283–287.

Kang G et al. (2001). Epidemiological and laboratory investigations of outbreaks of diarrhoea in rural South India: implications for control of disease. *Epidemiology and Infection*, 127:107–112.

Kemper KE (2004). Ground water – from development to management. *Hydrogeology Journal*, 12(1):3–5.

Kirshner AKT et al. (2008). Rapid growth of planktonic vibrio cholerae non-O1/non-O139 strains in a large alkaline lake in Austria: dependence on temperature and dissolved organic carbon quality. *Applied and Environmental Microbiology*, 74:2004–2015.

Kistemann T et al. (2002). Microbial load of drinking-water reservoir tributaries during extreme rainfall and runoff. *Applied and Environmental Microbiology*, 68(5):2188–2197.

Kovats RS, Hajat S, Wilkinson P (2004). Contrasting patterns of mortality and hospital admissions during hot weather and heat waves in Greater London, United Kingdom. *Occupational and Environmental Medicine*, 61:893–898.

Lake I et al. (2005). Effects of weather and river flow on cryptosporidiosis. *Journal of Water and Health*, 3:469–474.

Laursen, E et al. (1994). Gastroenteritis: a waterborne outbreak affecting 1600 people in a small Danish town. *Journal of Epidemiology and Community Health*, 48(5):453–458.

Lipp EK, Huq A, Colwell RR (2002). Effects of global climate on infectious disease: the cholera model. *Clinical Microbiology Reviews*, 15:757–770.

Lozano-Leon A et al. (2003). Identification of tdh-positive vibrio parahaemolyticus from an outbreak associated with raw oyster consumption in Spain. *FEMS Microbiology Letters*, 226:281–284.

Lucentini L et al. (2009). Unprecedented cyanobacterial bloom and microcystin production in a drinking-water reservoir in the south of Italy: a model for emergency response and risk management. In: Caciolli S, Gemma S, Lucentini L, eds. *Scientific symposium. International meeting on health and environment: challenges for the future. Abstract book*. Rome, Istituto Superiore di Sanità, 9–11 December (ISTISAN Congressi 09/C12) (http://www.iss.it/binary/imhe/cont/IMHE_Book_of_Abstracts_09_C12.pdf, accessed 6 July 2010).

Marangani J (2008). Proposal of a methodology for the optimal design of monitoring networks coastal aquifers management. In: *Proceedings of the 20th Salt Water Intrusion Meeting Program and proceedings book*. Naples, FL, 23–27 June:145–148 (http://www.conference.ifas.ufl.edu/swim/papers.pdf, accessed 14 September 2010).

McMichael AJ et al. (2004). Climate change and human health: present and future risks. *Lancet*, 367(9513):859–869.

Meehl GA et al. (2007). Global climate projections. In: Solomon S et al., eds. *Climate change 2007: the physical science basis. Contribution of Working Group I to the Fourth Assessment Report of the Intergovernmental Panel on Climate Change*. Cambridge, Cambridge University Press.

Meier HEM, Kjellström E, Graham LP (2006). Estimating uncertainties of projected Baltic Sea salinity in the late 21st century. *Geophysical Research Letters*, 33(15):L15705.

Menne B, Bertollini R (2000). The health impacts of desertification and drought. *Down Earth*, 14:4–6.

Menne B et al., eds. (2008). *Protecting health in Europe from climate change*. Copenhagen, WHO Regional office for Europe.

Meusel D et al. (2004). Public health responses to extreme weather and climate events – a brief summary of the WHO meeting on this topic in Bratislava on 9–10 February 2004. *Journal of Public Health*, 12(6):371.

Miettinen IT et al. (2001). Waterborne epidemics in Finland in 1998–1999. *Water Science and Technology*, 43:67–71.

Nchito M et al. (1998). Baboo. Cryptosporidiosis in urban Zambian children: an analysis of risk factors. *The American Journal of Tropical Medicine and Hygiene*, 59:435–437.

NDMC (2006). What is drought? [web site]. Lincoln, NE, University of Nebraska-Lincoln School of Natural Resources National Drought Mitigation Centre (http://www.drought.unl.edu/whatis/concept.htm#concept, accessed 14 September 2010).

OECD (2005). *Financing water supply and sanitation in eastern Europe, Caucasus and Central Asia. Proceedings from a conference of EECCA Ministers of Economy/Finance and Environment and their partners*. (Yerevan, 17–18 November) Paris, Organisation for Economic Co-operation and Development (http://www.oecd.org/dataoecd/29/46/36388760.pdf, accessed 5 April 2010).

OzCoast (2010). Saline intrusion [web site]. Canberra, ACT, OzCoast Australian Online Coastal Information (http://www.ozcoasts.org.au/indicators/saline_intrusion.jsp, accessed 9 July 2010).

PAHO (2000). *Proceedings of the 126th session of PAHO's executive committee*. Washington, DC, 26–30 June, Pan-American Health Organization/World Health Organization (http://www.paho.org/spanish/gov/ce/ce126_02.pdf, accessed 14 September 2010).

Pardue J et al. (2005). Chemical and microbiological parameters in New Orleans floodwater following Hurricane Katrina. *Environmental Science & Technology*, 39:8591–8599.

Parry ML et al., eds. (2007). *Climate change 2007: impacts, adaptation and vulnerability. Contribution of Working Group II to the Fourth Assessment Report of the Intergovernmental Panel on Climate Change (IPCC)*. Cambridge, Cambridge University Press.

Paz S et al. (2007). Climate change and the emergence of *vibrio vulnificus* disease in Israel. *Environmental Research*, 103(3):390–396.

PESETA (2008). The PESETA project – impacts of climate change in Europe [web site]. Brussels, European Commission Projection of Economic impacts of climate change in Sectors of the European Union based on bottom-up Analysis (PESETA) Project. (http://peseta.jrc.ec.europa.eu/, accessed 24 June 2010).

Phillippart CJM (2007). *Impacts of climate change on the European marine and coastal environment*. Strasbourg, European Science Foundation (Marine Board Position Paper 9) (http://peseta.jrc.ec.europa.eu/, accessed 26 June 2010).

Pond K et al. (unpublished). Health effects of climate change. In: Menne B et al. *Final report of the climate, environment and health action plan and information system project*. Copenhagen, WHO Regional Office for Europe.

Potter KW (1987). Research on flood frequency analysis: 1983-86. *Reviews of Geophysics*, 25(2):113–118.

Reacher M et al. (2004). Health impacts of flooding in Lewes: a comparison of reported gastrointestinal and other illness and mental health in flooded and non-flooded households. *Communicable Disease and Public Health*, 7:1–8.

Risebro HL et al. (2007). Fault tree analysis of the causes of waterborne outbreaks. *Journal of Water and Health*, 5(Suppl. 1):1–18.

Robine JM et al. (2008). Death toll exceeded 70 000 in Europe during the summer of 2003. *Comptes Rendus Biologies*, 331(2):171–178.

Saker ML et al. (2003). First report and toxicological assessment of the cyanobacterium cylindrospermopsis raciborskii from Portuguese freshwater. *Ecotoxicology and Environmental Safety*, 55(2):243–250.

Schijven JF, Hassanizadeh SM (2000). Removal of viruses by soil passage: overview of modelling, processes and parameters. *Critical Reviews in Environmental Science and Technology*, 31:49–125.

Schijven JF, De Rosa Husman AM (2005). Effects of climate changes on waterborne disease in the Netherlands. *Water Science and Technology*, 51:79–87.

Schmitz-Esser S et al. (2008). Diversity of Bacterial Endosymbionts of Environmental Acanthamoeba Isolates. *Applied and Environmental Microbiology*, 74(18):5822–5831.

Senhorst HAJ, Zwolsman JJG (2005). Climate change and effects on water quality: a first impression. *Water Science and Technology*, 51(5):53–59.

Sivonen K, Jones G (1999). Cyanobacterial toxins. In: Chorus I, Bartram J, eds. *Toxic cyanobacteria in water: a guide to their *public health consequences, monitoring and management*. London, E & FN Spon:41–111.

Spatharis S et al. (2007). Effects of pulsed nutrient inputs on phytoplankton assemblage structure and blooms in an enclosed coastal area. *Estuarine, Coastal and Shelf Science*, 73(3–4):807–815.

Sullivan CA, Meigh JR (2005). Targeting attention on local vulnerabilities using an integrated indicator approach: the example of the climate vulnerability index. *Water Science and Technology, Special Issue on Climate Change*, 51(5):69–78 (http://www.ncbi.nlm.nih.gov/pubmed/15918360, accessed 28 April 2010).

Sullivan CA, Huntingford C (2009). Water resources, climate change and human vulnerability. *Proceedings of the 18th World IMACS/MODSIM Congress, Cairns, 13–17 July* (http://www.mssanz.org.au/modsim09/I13/sullivan_ca.pdf, accessed 5 April 2010).

Swiss Confederation National Platform for Natural Hazards (2001). The cycle of integrated risk management [web site]. Bern, Swiss Confederation (http://www.planat.ch/index.php?userhash=47851918&l=e&navID=5, accessed 28 April 2010).

Thornton JA et al., eds. (1999). *Assessment and control of non-point source pollution of aquatic systems; a practical approach*. Paris and Carnforth, United Nations Educational, Scientific and Cultural Organization and Parthenon Publishing (Man and the Biosphere Series Volume 23).

Tison D et al. (1980). Growth of legionella pneumophila in association with blue-green algae (cyanobacteria). *Applied and Environmental Microbiology*, 39:456–459.

UNECE (2009a). *Convention on the Protection and Use of Transboundary Watercourses and International Lakes. Guidance on water and adaptation to climate change*. Geneva, United Nations (http://www.unece.org/env/water/publications/documents/Guidance_water_climate.pdf, accessed 5 April 2010).

UNECE (2009b). *Transboundary flood risk management: experiences from the UNECE region*. Geneva, United Nations.

UNHCR (2007). *Handbook for emergencies, 3rd edition*. Geneva, United Nations High Commission for Refugees (http://www.unhcr.org/471db4c92.html, accessed 28 April 2010).

UNISDR (2005). *Hyogo framework for action 2005–2015: building the resilience of nations and communities to disasters*. Geneva, United Nations International Strategy for Disaster Reduction (http://www.unisdr.org/wcdr/intergover/official-doc/L-docs/Hyogo-framework-for-action-english.pdf, accessed 28 April 2010).

UNISDR (2009). *UNISDR terminology on disaster risk reduction*. Geneva, United Nations International Strategy for Disaster Reduction (http://www.unisdr.org/eng/library/UNISDR-terminology-2009-eng.pdf, accessed 24 June 2010).

UNISDR (2010). Platform for the promotion of early warning [web site]. Geneva, United Nations International Strategy for Disaster Reduction (http://www.unisdr.org/ppew/whats-ew/basics-ew.htm, accessed 6 July 2010).

UNOCHA (2000). *United Nations disaster assessment and coordination (UNDAC) field handbook, 3rd edition*. Geneva, Office for the Coordination of Humanitarian Affairs Field Coordination Support Unit (http://www.reliefweb.int/undac/documents/UNDACHandbook.pdf, accessed 6 July 2010).

UNSW (2010). Potential impacts of sea-level rise and climate change in coastal aquifers [web site]. Sydney, NSW, University of New South Wales (http://www.connectedwaters.unsw.edu.au/resources/articles/coastal_aquifers.html, accessed 9 July 2010).

USDA (1954). US Salinity Laboratory. Diagnoses and improvement of saline and alkali soils. Agriculture Handbook, No. 60. In: Tanki KK, ed. (1990). *Agricultural salinity assessment and management*. Baltimore, MD, American Society of Civil Engineers (ASCE Manuals & Reports on Engineering Practice, No. 71).

Valent F et al. (2004). Burden of disease attributable to selected environmental factors and injury among children and adolescents in Europe. *Lancet*, 363(9426):2032–2039.

Vasconcelos P (2006). Flooding in Europe: a brief review of the health risks. *Eurosurveillance*, 11(4).

Wade TJ et al. (2004). Did a severe flood in the Midwest cause an increase in the incidence of gastrointestinal symptoms? *American Journal of Epidemiology*, 159:398–405.

Wasmund N, Uhlig S (2003). Phytoplankton trends in the Baltic Sea. *ICES Journal of Marine Science*, 60(2):177–186.

WBCSD (2008). *Adaptation – an issue brief for business*. Geneva, World Business Council on Sustainable Development (http://www.wbcsd.org/DocRoot/iMn5EtG4bkjxQNLfU9UZ/Adaptation.pdf, accessed 6 July 2010).

WHO (2003). *Emerging issues in water and infectious disease*. Geneva, World Health Organization.

WHO (2005). *Minimum water quantity needed for domestic use in emergency*. Geneva, World Health Organization (http://www.who.int/water_sanitation_health/hygiene/envsan/minimumquantity.pdf, accessed 5 April 2010).

WHO (2006). *Guidelines for drinking-water quality*. Geneva, World Health Organization (http://www.who.int/water_sanitation_health/dwq/gdwq3rev/en/index.html, accessed 6 July 2010).

WHO (2007). Legionella *and the prevention of legionellosis*. Geneva, World Health Organization (http://www.who.int/water_sanitation_health/emerging/leginella_rel/en/, accessed 5 August 2011).

WHO (2010). *Vision 2030 – the resilience of water supply and sanitation in the face of climate change*. Geneva, World Health Organization (http://www.who.int/water_sanitation_health/publications/9789241598422_cdrom/en/index.html, accessed 5 April 2010).

WHO Regional Office for Europe (2005). *Health and climate change: the now and how. A policy action guide.* Copenhagen, WHO Regional Office for Europe.

WHO Regional Office for Europe (2007). *Children's health and the environment in Europe. A baseline assessment.* Copenhagen, WHO Regional Office for Europe.

WHO Regional Office for Europe (2009). *Wastewater treatment and access to improved sanitation.* Copenhagen, WHO Regional Office for Europe (ENHIS 2007 fact sheet 1.3) (http://www.euro.who.int/__data/assets/pdf_file/0003/97365/ENHIS_Factsheet_1_3.pdf, accessed 23 March 2011).

WHO Regional Office for Europe (2011). *Small-scale water supplies in the pan-European region.* Copenhagen, WHO Regional Office for Europe.

WHO/UNICEF (2008). *Joint monitoring programme for water supply and sanitation (JMP): special focus on sanitation.* New York, NY and Geneva, United Nations Children's Fund and World Health Organization (http://www.who.int/water_sanitation_health/monitoring/contents.pdf, accessed 15 September 2010).

WHO/UNICEF (2010). *Joint monitoring programme for water supply and sanitation (2010). Progress on sanitation and drinking-water.* New York, NY and Geneva, United Nations Children's Fund and World Health Organization (http://www.who.int/water_sanitation_health/publications/9789241563956/en/index.html, accessed 16 September 2010).

Wittman RJ, Flick GJ (1995). Microbial contamination of shellfish – prevalence, risk to human health, and control strategies. *Annual Review of Public Health,* 16:123–140.

WMO (1989). *Statistical distributions for flood frequency analysis.* Geneva, World Meteorological Organization (WMO No. 718, Operational Hydrology Report No. 33).

WMO (2005). *Commission for basic systems, thirteenth session – abridged final report with resolutions and recommendations.* Geneva, World Meteorological Organization (WMO No. 985).

WMO (2006). *Comprehensive risk assessment of natural hazards.* Geneva, World Meteorological Organization (WMO/TD 955) (reprinted 2006) (http://www.wmo.int/pages/prog/drr/publications/drrPublications/TD0955_Comprehensive_Assessment_of_Natural_Hazards/WMO_TD0955e.pdf, accessed 16 September 2010).

WMO (2008). *Guide to meteorological instruments and methods of observation, 7th edition.* Geneva, World Meteorological Organization (WMO No. 8) (http://www.wmo.int/pages/prog/www/IMOP/publications/CIMO-Guide/CIMO_Guide-7th_Edition-2008.html, accessed 5 April 2010).

WMO (2009). *Guide to hydrological practices, 6th edition. Vol. I Hydrology – from measurement to hydrological information. Vol. II Management of water resources and application of hydrological practices.* Geneva, World Meteorological Organization (WMO No. 168).

Wolf T, Menne B, eds. (2007). *Environment and health risks from climate change and variability in Italy.* Copenhagen and Rome, WHO Regional Office for Europe and APAT Agency for Environment Protection and Technical Services.

Woodruff RE et al. (2002). Predicting Ross River virus epidemics from regional weather data. *Epidemiology,* 13:384–393.

World Bank (2006). *Drought: management and mitigation assessment for Central Asia and the Caucasus – regional and country profiles and strategies.* Washington, DC, World Bank.

Bibliography

A

Apfel F et al. (2010). *Health literacy part 2: "Evidence and case studies".* Axbridge, World Health Communications Associates Ltd.

Anon (2010). *Promoting health – advocacy guide for health professionals.* Axbridge, World Health Communications Associates Ltd.

APFM (2007). *Economic aspects of integrated flood management.* Geneva, Associated Programme on Flood Management (WMO No. 1010).

B

Bartram J et al. (2009). *Water safety plan manual: step-by-step risk management for drinking-water suppliers.* Geneva, World Health Organization/International Water Association (http://www.who.int/water_sanitation_health/publication_9789241562638/en/index.html, accessed 5 April 2010).

Boost M et al. (2008). Detection of acanthamoeba in tap water and contact lens cases using polymerase chain reaction. *Optometry and Vision Science*, 85(7):526–530.

Bouma MJ, Dye C, Van der Kaay HJ (1996). Falciparum malaria and climate change in the Northwest Frontier Province of Pakistan. *American Journal of Tropical Medicine Hygiene*, 55(2):131–137.

Bouma MJ, Dye C (1997). Cycles of malaria associated with El Niño in Venezuela. JAMA, 278(21):1772–1774.

D

De Graf RE, Van de Ven FHM. *Transitions to more sustainable concepts of urban water management and water supply. 10th International Conference on Urban Drainage.* Copenhagen, 21–26 August.

K

Kos M (2003). Povodně a ČOV [Floods and wastewater treatment plants]. In: Dián M, ed. *Rekonštrukcie stokových sietí a čistiarní odpadových vod [Reconstructions of sewerage systems and wastewater treatment plants]*. Liptovský Ján, 27 October. Bratislava, VÚVH:150–158.

Kravčík M et al., eds. (2008). *Water for the recovery of the climate – a new water paradigm*. Košice, Typopress s.r.o (http://www.waterparadigm.org, accessed 5 July 2010).

M

Ministry of the Environment (2004). *Výsledná zpráva projektu Vyhodnocení katastrofální povodně v srpnu 2002 a návrhu úpravy systému prevence před povodněmi. [Assessment of the catastrophic flood in August 2002 and proposal for adaptation of the flood prevention system – final report of the project]*. Prague, Ministerstvo životního prostředí ČR. Ministry of Environment (of the Czech Republic).

Mubareka S et al. (2006). Acanthamoeba species keratitis in a soft contact lens wearer molecularly linked to well water. *The Canadian Journal of Infectious Diseases & Medical Microbiology*, 17(2):120–122.

Munich Re (2009). *Highs and lows: weather risks in central Europe*. Munich, Munich Re.

Mutňanský A, Neužil J (2003). Zkušenosti z likvidací povodňových škod na komplexu ČOV v Roztokách [Experience from the flood damage cleanup in the UWWTP in Roztoky]. In: Dián M, ed. *Rekonštrukcie stokových sietí a čistiarní odpadových vod [Reconstructions of sewerage systems and wastewater treatment plants]*. Liptovský Ján, 27 October. Bratislava, VÚVH:175–183.

R

Remešová Ž et al. (2003). Poznatky z uvádění biologických ČOV zasažených povodní do provozu [Findings of putting flooded biological WWTPs into operation]. In: Dián M, ed. *Rekonštrukcie stokových sietí a čistiarní odpadových vod [Reconstructions of sewerage systems and wastewater treatment plants]*. Liptovský Ján, 27 October. Bratislava, VÚVH:160–167.

S

SIWI (2008). *Progress and prospects on water: for a clean and healthy world with special focus on sanitation*. Stockholm, Stockholm International Water Institute (http://www.siwi.org/documents/Resources/Synthesis/Synthesis_full_version_08.pdf, accessed 5 April 2010).

U

UNISDR (2005). *Final report of the world conference on disaster reduction*. New York, NY, United Nations (A/CONF.206/6).

V

Visvesvara GS et al. (2007). In vitro culture, serologic and molecular analysis of acanthamoeba isolated from the liver of a keel-billed toucan (ramphastos sulfuratus). *Veterinary Parasitology*, 143(1):74–78.

W

WHO Regional Office for Europe (2011). Environment and Health Information System (ENHIS) [web site]. Copenhagen, WHO Regional Office for Europe (http://www.euro.who.int/en/what-we-do/data-and-evidence/environment-and-health-information-system-enhis, accessed 23 March 2011).

WMO (2007). *Economic aspects of integrated flood management*. Geneva, World Meteorological Organization Associated Programme on Flood Management (APFM) (http://www.apfm.info/pdf/ifm_economic_aspects.pdf, accessed 6 July 2010).

Z

Zábranská J, Dohanyos M (2000). Obnovení provozu kalového hospodářství na ÚČOV Praha [Restoration of sludge management in the central UWWTP of Prague.]. In: Dián M, ed. *Rekonštrukcie stokových sietí a čistiarní odpadových vod [Reconstructions of sewerage systems and wastewater treatment plants]*. Liptovský Ján, 27 October. Bratislava, VÚVH:168–174.